The Hope of Progress

First published in 1972, *The Hope of Progress* presents collection of essays and lectures dealing with the history of scientific ideas and the impact of science on society. The principle piece in this volume is the author's 1969 presidential address to the British Association 'On The Effecting of All Things Possible', an argument for believing in the ability of science to solve the problems it has itself created, and which too many of us believe insoluble. It contains author's Romanes Lecture on 'Science and Literature' and a well known critique of J.D. Watson's notorious account of the discovery of the molecular structure of DNA, *The Double Helix*. Other chapters discuss the possibility of the control and domination by science of the body and mind of Man- though the author concludes in 'The Genetic Improvement of Man' : 'I think that, in the main, for many centuries to come, we shall have to put up with human beings as they are at present constituted'. This book will be useful for scholars and researchers of history of science, philosophy of science, natural science, and philosophy in general.

The Hope of Progress

by P.B. Medawar

Routledge
Taylor & Francis Group

First published in 1972
By Methuen & Co.

This edition first published in 2021 by Routledge
2 Park Square, Milton Park, Abingdon, Oxon, OX14 4RN
and by Routledge
605 Third Avenue, New York, NY 10017

Routledge is an imprint of the Taylor & Francis Group, an informa business

Publisher's Note
The publisher has gone to great lengths to ensure the quality of this reprint but points out that some imperfections in the original copies may be apparent.

Disclaimer
The publisher has made every effort to trace copyright holders and welcomes correspondence from those they have been unable to contact.

A Library of Congress record exists under LCCN: 72184400

ISBN 13: 978-1-032-11806-2 (hbk)
ISBN 13: 978-1-003-22161-6 (ebk)
ISBN 13: 978-1-032-11807-9 (pbk)

DOI: 10.4324/9781003221616

The Hope of Progress

P. B. MEDAWAR, F.R.S.

METHUEN & CO LTD

11 New Fetter Lane, London EC4

First published 1972 by
Methuen & Co Ltd
© *1972 by P. B. Medawar*
Printed in Great Britain by
Cox & Wyman Ltd
Fakenham, Norfolk

SBN 416 65720 6

To our children – and theirs

Contents

Introduction

The title of this collection comes from a paragraph towards the end (p. 127) of my Presidential Address in 1969 to the British Association 'On "The Effecting of all Things Possible"'.

> To deride the hope of progress is the ultimate fatuity, the last word in poverty of spirit and meanness of mind. There is no need to be dismayed by the fact that we cannot yet envisage a definitive solution of our problems, a resting-place beyond which we need not try to go . . .

In his annual Address to the British Association the President speaks *of* and *for* science. His remarks sometimes give the impression of a moralizing discourse or sermon and I took no particular pains to avoid giving this impression. Like many other sermons it has a sanguine temper and a cheerfulness for which there may seem to be no reasonable grounds – a point I shall return to later in this Introduction. I have not rewritten the other pieces to conceal the fact that most of them began as Lectures.

'Science and Literature' was the Romanes Lecture delivered in Oxford in 1968 and published in *Encounter* in January 1969. After the publication of the Lecture the Editor of *Encounter*, Mr Lasky, invited Mr John Holloway, the poet and critic, to comment upon it. This led to a civilized exchange of views which was also published in *Encounter*, July 1969. I am grateful to Mr Holloway for allowing his contribution to the duologue to be reproduced here. I have only one clarifying comment to make. I do not take all criticism to be equivalent to literary criticism. I believe that criticism in some more general sense has a very important function to play in inquiring into the motives and purposes of scientists and the direction towards which their work is leading.

Mr Lasky also invited the psychologist Dr Anthony Storr to comment on the depreciatory remarks about psychoanalysis in my Romanes Lecture. Dr Storr's criticisms were duly published, but he did not know that I was to be invited to write a rejoinder to them, so he understandably preferred that his commentary should not be republished here. Instead I have published a short article ('Further Comments on Psychoanalysis') which is based essentially on my reply to Dr Storr. It deals with a problem in psychology which has something in common with the old nature/nurture problem in genetics. It is the problem of the degree to which disturbances of behaviour may be attributed either to purely behavioural or psychogenic influences or to physical abnormalities in the structure or physical working of the brain.

'The Genetic Improvement of Man' (pp. 69–76) was my contribution to a symposium honouring the great Australian biologist Sir Macfarlane Burnet. Its purpose was to undermine an argument often used by so-called 'positive' eugenicists, that the calculated breeding of domesticated animals, particularly of dogs and horses, shows that it should in principle be possible to improve the quality of mankind, even to superman level, by a regimen of directed mating as Campanella and other Utopian thinkers have hinted. The disagreeable element of this procedure, culling, in effect the killing off of unwanted variants, is conveniently forgotten. I believe this scheme to be quite mistaken, and thought it important to explain why. When I wrote this article I had not had the pleasure of reading Professor R. D. Passmore's admirable *The Perfectibility of Man* (London, 1970), a book which deals much more thoroughly with the notion of perfectibility in general. The distinction I draw between Arcadian and Utopian notions or day dreams of a better world still strikes me as useful, but I am far from satisfied with what I call the 'Olympian' notion or anyhow with my account of it. Passmore's book fully remedies this deficiency. The most famous literary exponent of Arcadian thought was, of course, Jean-Jacques Rousseau. Fortunately for him, he was not called upon to translate his ideas into terms of real life.

The essay on Animal Experimentation represents for the year 1966 the Research Defence Society's annual attempt to counteract the propaganda of those who would like to prohibit the use of animals in biological and medical research, perhaps because the welfare of animals genuinely means more to them than the welfare of human beings. There are two points I should like specially to emphasize. The first is the strength of the biologist's ambition to replace the use of animals by purely chemical or cellular systems. The second is how little the general public are to be relied upon as judges of animal welfare. The history of the breeding in this country of fancy or curiously abnormal dogs is a most discreditable one and we must not forget that the problems it gives rise to are a direct consequence of the patronage and taste of the general public.

'Science and the Sanctity of Life' (in *Life or Death: Ethics and Options* published by the University of Washington Press, 1968) was my contribution to a Symposium at Reed College, Portland, Oregon, which had to do with the various ways in which science or technology might devalue certain elementary human rights and privileges. 'The Genetic Improvement of Man' overlaps with but amplifies one passage in the former which may be thought rather obscure, that which deals with the case against positive eugenics.

'Lucky Jim' (not my title) was my critique for the *New York Review of Books* of J. D. Watson's notorious account (*The Double Helix*) of the discovery of the molecular structure of DNA – the substance which, being responsible for the continuity of life, has some claim to be considered the most important of all chemical compounds. No passage in Watson's book caused more offence than that which described his absolute determination to get in first with the solution of his problem, especially as Linus Pauling was only a head behind him. Here I defended Watson, for the reasons I give. It just so happens that a few months ago I learned that an ingenious American experimental surgeon had put forward a hypothesis I myself should have liked to have been the first to propound. This gave me an opportunity to examine my reactions to a situation similar in principle to that which confronted Watson, though the problem was nothing like so important. I was relieved to find that my first reaction was one of genuine admiration for him, combined with a pleasant feeling of reassurance that my colleagues and I were not going to be alone in studying a rather difficult problem that might lead nowhere. The element of camaraderie or mateyness in scientific research is one of the attractions of the scientific life. It was perceptive of Mr Holloway, in his critique of my Romanes Lecture (p. 39), to call attention to the 'great collaborative expertise' of science – that which gives the authoritative plural form to 'we think that ...' and 'we no longer believe ...' – it wouldn't last very long if philosophies like J. D. Watson's became at all common.

The theme of my BA Address was that there can be technological remedies for technological evils, even for the greatest evil of all – the population problem, the magnitude and gravity of which are now pretty well understood. The population problem is a direct consequence of the advances in medicine and sanitary engineering over the past hundred years which have greatly reduced human mortality, particularly in infancy and childhood. Nevertheless, I believe that a medically and otherwise unobjectionable method of preventing conception will one day be devised. The introduction of such a method into use is naturally beset by grave political, administrative, and doctrinal problems. But these problems are soluble and must be solved.

Anybody who denies that a technological remedy is possible for a technological evil is making a prediction of just the kind – that a certain procedure is inconceivable or impossible – that history most often and most easily falsifies. Surely a visit to the other side of the moon has convinced us that almost any material accomplishment that does not violate the laws of physics is within our power, given an unswerving determination to achieve it, which in turn implies caring enough about the consequences of failure. It was some such thought I had in mind when I decided upon a Baconian motto ('the effecting of all things possible') for my Address to the British Association.

It was because of the prevailing attitudes of despondency and hopelessness about our present predicament, that I drew a parallel between the climates of opinion in the seventeenth century and today, a parallel which, I see, has also been drawn by Chomsky,[1] though on somewhat different grounds. Today there is a perverse willingness to believe that the advances of science and technology bring with them some *essential* malefaction. This view can only be sustained by systematically disregarding all the benefactions of science, and thinking only of its harmful effects, real or imagined. People in this mood seem almost to welcome evidence of some new ecological miscarriage or misadventure of technological origin. It confirms them in their expectation of imminent doom. In this attitude there is also an element of cultural counter-attack, from some of the relict faunas of Oxford high tables, the people who never did believe in science, and knew that no good would come of it. Another very understandable reason for fearfulness is that laymen simply do not realize the width of the gap between showing that a certain scheme or advance is conceivable (as Leonardo conceived the possibility of a submarine vessel) and causing it to come about in practice. The existence of this large gap makes it possible to control science far more tightly than most people realize. I shall illustrate this point by reference to research on aging, aspects of which have been the subject of a discussion, in *The Times* (27 February 1971) and on the BBC, in which a former Minister for Science took part. Research is in progress at the National Institute for Medical Research which might ultimately have the effect of extending the lifespan of certain simple fungi for a few hours or days or of giving tissues cultured outside the body a few extra cell divisions of life. Fearful laymen at once translated this into a scheme to populate an already overburdened world with centenarians kept alive by scientific tricks. Here the gap between the workbench and the demographic disaster is to be measured in tens of thousands of man-hours and hundreds of thousands of pounds. Where are these large sums to come from? There are no vacancies in the Medical Research

[1] N. Chomsky, *Language and Mind* (New York, 1968).

Council or in most University Medical Schools for research workers with socially disruptive ambitions. As for the money, no one will expect it from the Medical Research Council, or from any of the great research foundations. Anyone who is still determined to believe the worst of scientific research will have to fall back upon those trusty characters of Gothic fiction: the mad scientist and the lunatic benefactor. If laymen realized how very difficult it is to find even sane benefactors for some biological research they would have less cause to worry over the consequences of mad patronage. Those who think the mad scientist a bit far-fetched should call to mind other no less extreme misapprehensions about the nature of science and scientists. Towards the end of the war a bishop wrote gravely to *The Times* inviting all nations to destroy 'the formula' of the atomic bomb. There is no simple remedy for ignorance so abysmal.

Some thinkers of the twentieth century believe that civilized society is the victim of a nervous ailment. A 'nervous breakdown' was J. M. Keynes's diagnosis, and 'a neurosis' Freud's.[1] Keynes believed that mankind has been 'expressly evolved by nature – with all our impulses and deepest instincts – for the purpose of solving the economic problem'. Now the economic problem was approaching solution – 'I think with dread', Keynes goes on to say, 'of the readjustment of the habits and instincts of the ordinary man bred into him for countless generations that he may be asked to discard within a few decades. To use the language of today must we not expect a general nervous breakdown?' This argument is a relic of social Darwinism. Freud's, on the other hand, is a relic of recapitulation theory, the now discredited view that there is a very exact parallel between the development of an individual and the evolution of a lineage, so that, in the usual formula, a developing animal 'climbs up its own family tree'. Freud reasons that the development of civilization is an evolutionary process and goes on to say 'If the development of civiliza-

---

[1] *Economic Possibilities for our Grandchildren*, J. M. Keynes, The Nation and Athenaeum, 18 October 1930. S. Freud, *Civilization and its Discontents*, ed. J. Strachey (Hogarth Press and Institute of Psychoanalysis, 1965).

tion has such a far-reaching similarity with the development of an individual and if it employs the same methods, should we not be justified in coming to the conclusion that in some civilizations or epochs of civilization possibly even the whole of mankind have become neurotic?'

These arguments are not convincing. The idea that civilized society can suffer from an ailment is a figure of speech which can easily lead us astray. It is fair enough to speak of society's suffering from a 'disorder' because orderliness is one of the defining characteristics of a civilized society. But we shall depress ourselves unduly if we think of ourselves as sitting by civilization's bedside. It really makes no more sense to speak of civilization's suffering a nervous breakdown than to speak of its having a stomach ache or a bad back. The disorders of society are peculiar to and distinctive of society, e.g. an unfavourable trade balance, unemployment or over-population. They are particular disorders, moreover, for which we should seek particular remedies. I feel that the misuse of the organismal conception of society must bear some part of the blame for the current feeling of hopelessness and despondency about how matters can be remedied.

Another figure of speech we may justifiably deplore is that of the Mastery of Nature or Subjugation of Nature. The concept is usually attributed to Bacon but in the context of Bacon's own thought it is not specially offensive, because it has a technical meaning connected with his advocacy of experimentation. The idea makes a brief appearance in the writings of Marx[1] and of Freud.[2] The conquest of infectious disease sounds perfectly all right, because bacteria are indeed inimical to us. On the other hand the 'conquest of space' rings quite false to me. The main objection to the ideology of mastery and warfare is that it dulls the sensibilities and seems to condone or in some perverse way

[1] See D. McLellan, 'Marx and the Missing Link', *Encounter*, November 1970, in which special reference is made to his forthcoming edition of selections from Marx's *Grundrisse* (New York, 1971).

[2] S. Freud, *Civilization and its Discontents*, ed. J. Strachey (Hogarth Press and Institute of Psychoanalysis, 1965).

to justify even the worst excesses of environmental despoliation. I hope that these unpleasant figures of speech have not taken so deep a hold that they cannot be uprooted from popular writing and popular thought. It is *understanding*, not mastery, that should be the ambition of scientific research.

Another factor which makes for uneasy relations between technology and civilized life is the tendency, especially prevalent in this country because of its alleged economic backwardness, to make economic return or cost effectiveness the ultimate measure of the worthwhileness of any enterprise.[1] The economic system of mensuration is not conducive to the welfare of the environment. The purification of toxic effluents and their safe disposal are costly obligations and the cost effectiveness of many manufacturing procedures is greatly improved when these obligations are skimped or disregarded.

I fully support Professor Paul Ehrlich's contention[2] that we shall have to stop working towards and making a virtue of an economy marked by profligate overproduction, consumption and waste and work towards something more like a space-ship economy of which the distinguishing features are frugality, recycling and above all forethought.

To go back finally to the theme of my BA Address: although I have been very much criticized for saying so, I still believe that scientific or technological remedies can be found for most of the hateful and unintended misadventures or miscarriages associated with the advance of technology. This is not to say that the remedies I have in mind are wholly or even mainly scientific. The discovery of how to decontaminate a certain toxic effluent is obviously useless unless legislative and administrative action between them make it obligatory that the remedy should be used. Perhaps I *am* saying that I have more faith in scientific than in

[1] I learned from Rashi Fein that a scientist, Sir William Petty, FRS, was the first to attempt to cost the value of human life in £ Sterling and to make this a basis for recommendations about health policy. The value he put on human life was high enough to justify his recommending the institution of lying-in hospitals for illegitimate children to be brought up to undertake fair and hard work.

[2] See *Encounter*, October and December 1970.

political man if only because the solution of the scientific element of the problem is much easier, being so much less handicapped by the inertia of bigotry and self-interest. In a cooperative society the race which Hobbes likened to life (p. 127) is a relay race in which the scientist can generally be relied upon to pass the baton to the next member of the team. It is not often he, I respectfully submit, who drops the baton or loses it altogether.

Putting together a number of published lectures and writing an overall introduction sounds an easy enough task, but in reality it was not so. My wife's help and encouragement were decisive in making it possible for this book to appear. I should like also to make special mention of the skill and patience of my secretarial staff.

I am very grateful to a number of editors and publishers for permission to reproduce articles or lectures that first appeared in their pages. *Encounter*: 'Science and Literature', 'Science and the Sanctity of Life' (and here I must also acknowledge the kindness of the Washington University Press). *The New York Review of Books*: 'Lucky Jim'. *The Australasian Annals of Medicine*: 'The Genetic Improvement of Man'.

B

Science and Literature

I

I hope I shall not be thought ungracious if I say at the outset that nothing on earth would have induced me to attend the kind of lecture you may think I am about to give. Science and literature – what a hackneyed subject, you must feel. Must we go into that again? What can there be to say that has not already been very well said by I. A. Richards, Aldous Huxley, C. P. Snow, Martin Green, J. Bronowski, D. G. King-Hele, and half a dozen others?[1]

Let me begin with an outline of some of the things I do *not* intend to say. I shall say nothing whatsoever about education, and have no formula for compounding science and literature into a single diet; nor shall I say, or even be thinking, that imaginative literature can be regarded as an antidote or counter-irritant to science, or vice versa. There will be no readings from poetry written by scientists – not even a quotation or biopsy specimen from the poetry of George John Romanes, FRS.[2] I shall not declare that henceforward the discoveries, ideas and adventures of science should become a bigger part of the subject matter of poetry, as Wordsworth[3] thought they might; nor shall I reproach poets and 'magazine critics', as Peacock did,[4] with carrying on

---

[1] I shall quote below from I. A. Richards, *Science and Poetry* (London, 1926); Aldous Huxley, *Literature and Science* (London, 1963). See also C. P. Snow, *The Two Cultures: And a Second Look* (Cambridge, 1964); Martin Green, *Science and the Shabby Curate of Poetry* (London, 1964); J. Bronowski, *Science and Human Values*, rev. ed. (New York, 1965); D. G. King-Hele, *Shelley: His Thought and Work* (London, 1960); *Erasmus Darwin* (London, 1963).

[2] See *A Selection from the Poems of George John Romanes*, ed. T. Herbert Warren (London, 1896). I am much obliged to Mr R. B. Freeman for calling my attention to Romanes's poetry. I have never read worse.

[3] See Wordsworth's introduction to the *Lyrical Ballads* (1802 edition).

[4] Peacock, *The Four Ages of Poetry*, 1820.

just as if 'there were no such things in existence as mathematicians, astronomers, chemists, moralists, metaphysicians, historians, politicians, and political economists'.

After these various abjurations, what is there left to say? If I had to choose a motto for this lecture, I should turn a remark of Lowes Dickinson's upside down. 'When science arrives,' said Lowes Dickinson,[1] 'it expels literature' – an echo, perhaps, of Keats's lament that science unweaves the rainbow and makes a dull ordinariness out of awful things. The case I shall find evidence for is that when literature arrives, it expels science. There are large territories of human belief and learning upon which both science and literature have very important things to say, for example, social and cultural anthropology, psychology and human behaviour generally, and even cosmology. These subjects lie within the compass of literature in so far as they have to do with human hopes, fears, beliefs and motives; with the attempt to give an account of ourselves and investigate our condition; and with matters of general culture, by which I mean the whole pattern of the way in which people think and carry on. The case that can be made for science is that in all these subjects, we have *also* to work towards a special kind of understanding which, though imaginative in origin (as I shall hope to convince you), is under the censorship or restrictive influence of a certain kind of obligation towards the truth.

The way things are at present, it is simply no good pretending that science and literature represent complementary and mutually sustaining endeavours to reach a common goal. On the contrary, where they might be expected to co-operate, they compete. I regret this very much, don't think it necessary, and wish it were otherwise. We are going through a bad episode in cultural history. We all want to be friends, and one day perhaps we shall be so. 'Let us advance together, men of letters and men of science,' said Aldous Huxley,[2] 'further and further into the ever expanding regions of the unknown.' That is a fine ambition, though most of

[1] Lowes Dickinson, *Plato and His Dialogues* (London, 1931).
[2] Huxley, *op. cit.* p. 18 n. 1.

us will feel awkward at its wording; but if it is to be achieved, scientists and men of letters must work their way towards an understanding – not just of each other's accomplishments (there is mischief and magnificence in both), nor just of each other's purposes (which are doubtless mixed, though officially both are good), but of each other's methods and energizing concepts and the quality and pattern of movement of each other's thought. I want therefore to discuss imagination and criticism as they enter into science on the one hand and into literature on the other; to explain why I think that scientific and literary conceptions of style and matters of communication cannot be reconciled; and finally to compare scientific and poetic notions of the truth. Towards the end, if there is time, I shall use Freudian psycho-analysis and existential psychiatry to illustrate the way in which science and literature compete for the territories on which they both have claims.

II

Let me begin by discussing the character and interaction of imagination and critical reasoning in literature and in science. I shall use 'imagination' in a modern sense (modern on the literary time scale, I mean), or, at all events, in a sense fully differentiated from mere fancy or whimsical inventiveness. (It is worth remembering that when the phrase 'creative imagination' is used today, we are expected to look solemn and attentive, but in the eighteenth century we could as readily have looked contemptuous or even shocked.)

The official Romantic view is that Reason and the Imagination are antithetical, or at best that they provide alternative pathways leading to the truth, the pathway of Reason being long and winding and stopping short of the summit, so that while Reason is breathing heavily, there is Imagination capering lightly up the hill. It is true that Shelley[1] recognized a poetical element in science, though 'the poetry is concealed by the accumulation of

[1] Percy Bysshe Shelley, *A Defence of Poetry* (1821).

facts and calculating processes'; true also that in one passage of his famous rhapsody, he was kind enough to say that poetry comprehends all science – though here, as he makes plain, he is using poetry in a general sense to stand for all exercises of the creative spirit, a sense that comprehends imaginative literature itself as one of its special instances. But in the ordinary usages to which I shall restrict myself, Reason and Imagination are antithetical. That was Shelley's view and Keats's, Wordsworth's, and Coleridge's; it was also Peacock's, for whom Reason was marching into territories formerly occupied by poets; and it was also the view of William Blake,[2] who came 'in the grandeur of Inspiration to cast off Rational Demonstration ... to cast off Bacon, Locke, & Newton'; 'I will not Reason & Compare – my business is to create.'

This was not only the official view of the Romantic poets; it was also the official scientific view. When Newton wrote *Hypotheses non fingo*, he was taken to mean that he reprobated the exercise of the imagination in science. (He did not 'really' mean this, of course, but the importance of his disclaimer lies precisely in this misunderstanding of it.) Bacon too, and later on John Stuart Mill were taken as official spokesmen for the belief that there existed, or could be devised, a calculus of discovery, a formulary of intellectual behaviour, which could be relied upon to conduct the scientist towards the truth, and this new calculus was thought of almost as an antidote to the imagination, as it had been in Bacon's own day an antidote to what Macaulay[2] called the 'sterile exuberance' of scholastic thought. Even today this central canon of inductivism – that scientific thought is fully accountable to reason – is assumed quite unthinkingly to be true. 'Science is a matter of disinterested observation, unprejudiced insight and experimentation, and patient ratiocination within some system of logically correlated concepts' – an important

---

[1] William Blake, *Milton* (1804), book 2, pl. 41; and *Jerusalem* (1804), chap. 1, pl. 10.

[2] Thomas Babington Macaulay, *Lord Bacon* (1837), an extended review of Montagu's edition of Bacon's works that first appeared in the *Edinburgh Review*.

opinion, for Aldous Huxley (see n. 1 p. 18) is a man thought to speak with equal authority for science and letters.

Huxley would, of course, have been the last man to deny imagination a role in science; nor even did the high priest of Inductivism, Karl Pearson; but in science a creative imagination is the privilege of the rare spirit who achieves in a blaze of intuition what the rest of us can only do by rote or by 'analytic industry' (Wordsworth's term). But the point is (I am still recounting the official view) that we *can* do it; we may not all be great cooks, but we can all read the instructions on the packet. There *is* a calculus of discovery, and it works independently of intuition, though nothing like so fast.

The reductionist view – of the *complete* accountability of science to reason – is no longer believed in by most people who have thought deeply about the nature of the scientific process. An entirely different conception grew up in the writings of William Whewell, Stanley Jevons, C. S. Peirce, and latterly of Karl Popper.[1] Because its message is in danger of being lost in technical discussion over points of detail, let me explain the gist of it in very general terms.

All advances of scientific understanding, at every level, begin with a speculative adventure, an imaginative preconception *of what might be true* – a preconception which always, and necessarily, goes a little way (sometimes a long way) beyond anything which we have logical or factual authority to believe in. It is the invention of a possible world, or of a tiny fraction of that world. The conjecture is then exposed to criticism to find out whether or not that imagined world is anything like the real one. Scientific reasoning is therefore at all levels an interaction between two episodes of thought – a dialogue between two voices, the one imaginative and the other critical; a dialogue, if you like, between the possible and the actual, between proposal and disposal, conjecture and criticism, between what might be true and what is in fact the case.

---

[1] For the appropriate references, consult my *The Art of the Soluble* (London, 1967) and *Induction and Intuition in Scientific Thought* (London, 1969).

In this conception of the scientific process, imagination and criticism are integrally combined. Imagination without criticism may burst out into a comic profusion of grandiose and silly notions. Critical reasoning, considered alone, is barren. The Romantics believed that poetry, *poiesis*, the creative exploit, was the very opposite of analytic reasoning, something lying far above the common transactions of reason with reality. And so they missed one of the very greatest of all discoveries, of the synergism between imagination and reasoning, between the inventive and the critical faculties.[1] I call it a 'discovery', but no one person made it. Coleridge could have made it; he alone in 150 years was qualified in every way to do so. It is a tragedy of cultural history that he did not.

At this point a spokesman for literature might say, 'I accept the idea that scientific reasoning can be resolved into a dialogue between critical and inventive faculties, or something of that general nature, but what is so distinctively scientific about it, and why should it be held to distinguish science from imaginative literature?' He would quote Matthew Arnold,[2] perhaps – all poetry is criticism of life' – but in this context he would probably not wish to press the point, for Arnold, too, saw criticism and inventiveness as antithetical, and when he says, for example, that 'the critical power is of lower rank than the creative', he shows that he has no idea of the existence of forms of intellection to which class distinctions of this kind do not apply. But would it not be reasonable to say that literary criticism has a function cognate with that which I have attributed to criticism in science?

I am not deeply enough read in modern critical literature to say

[1] After I delivered this lecture, Sir Ewart Jones called my attention to J. H. Van't Hoff's highly original inaugural lecture on 'Imagination in Science', given in 1878 when he was twenty-six years old. Van't Hoff speaks of the 'synergism of imagination and critical judgment', though not quite in the sense I intended here. His lecture has recently been translated and annotated by Professor G. F. Springer (Berlin, 1967).

[2] Matthew Arnold, *The Study of Poetry* and *The Function of Criticism at the Present Time*, reprinted in *Essays in Criticism*, first series (1865) and second series (1888).

whether there is anything in such an argument or not, but my inclination would be to say that there is not. I have been saying that the critical episode of thought is something integral with scientific reasoning; that which has not yet been exposed to it is not yet science. Literary criticism, on the other hand, is a branch of literature which has literature as its subject matter, and that is an altogether different thing. A better case could be made for saying that literary criticism has something in common with scientific methodology. There is something in this, surely, but the similarities between them are rather dull and the differences interesting. Scientific methodology has to do with matters of validation and justification as they enter into all forms of scientific thinking, with the trustworthiness of evidence, and with the analysis of certain formal ideas that are common to all the sciences, for example, the ideas of causality or of reducibility and emergence. But it has nothing to do with the motives and purposes of scientists or with the degree to which their work achieves them: science is known to us in terms of accomplishment, not in terms of endeavour. It does not attempt to justify science in any sense except the scientific; above all it does not try to see scientific thought and action as elements of general culture. What *should* be the equivalent in science of literary criticism is therefore represented by a great emptiness which is a reproach to all scholars, scientists and humanists alike. I cannot even think of a name for the new discipline that might fill those empty spaces, for 'scientific humanism' is too deeply committed to a different meaning, and its practitioner, the 'scientific humanist', has too long and too often been made a figure of fun. (You have only to think of some of Peacock's caricatures, or of Sir Austin Feverel.)[1] Perhaps the word I want is 'criticism' itself, without qualification.

The gist of what I have been saying is this. Our traditional views about imagination and criticism in literature and in science are based upon the literary propaganda of the Romantic poets and

---

[1] In George Meredith's *The Ordeal of Richard Feverel* (1859), the tiresome Sir Austin is explicitly described as a 'scientific humanist', the first example of this particular usage I have come across.

the erroneous opinions of inductivist philosophers. Imagination is the energizing force of science as well as of poetry, but in science imagination and a critical evaluation of its products are integrally combined. To adopt a conciliatory attitude, let us say that science is that form of poetry (going back now to its classical and more general sense) in which reason and imagination act together synergistically. This simple formal property (which can, of course, be set out in a much more professional and specific language than anything I have attempted here) represents the most important methodological discovery of modern thought.

III

I now turn to a discussion of matters of style.

The poet 'yieldeth to the powers of the mind an image of that whereof the philosopher bestoweth but a wordish description, which doth neither strike, pierce, nor possess the sight of the soul'.[1] At a time when the writing of English had reached a peak of adventurousness and effulgence – and partly, but not wholly, because it had done so – the New Philosophers of the seventeenth century (new scientists, we should now say) were put officially on their guard against the danger of being carried away by the sound of their own voices. I say 'officially' because the warning came from the Royal Society. Then and for evermore they were to abjure the 'painted scenes and pageants of the brain'.[2] Their writing was 'manly and yet plain ... It is not broken by ends of Latin, nor impertinent quotations, ... not rendered intricate by long parentheses, nor gaudy by flaunting metaphors; not tedious by wide fetches and circumferences of speech.' The scientific style was to be 'as polite and as fast as marble'. I am quoting Joseph Glanvill, FRS[3] – not a good one to talk, perhaps, as his own style was described by H. Oldenburg as somewhat florid; but he

---

[1] Sir Philip Sidney, *An Apology for Poetry* (1595).

[2] Abraham Cowley, *To the Royal Society* (1663).

[3] Joseph Glanvill, *Plus Ultra* (1668). I learned Henry Oldenburg's opinion from Jackson Cope's introduction to a recent facsimile edition (Gainesville, 1958).

spoke for common opinion. Not Words but Works were to carry the message of the New Philosophy. 'We believe a scientist because he can substantiate his remarks,' said I. A. Richards, 'not because he is eloquent and forcible in his enunciation. In fact, we distrust him when he seems to be influencing us by his manner.'[1] There is a passage in a still undiscovered manuscript of Bacon's in which he expresses his abhorrence of the *venditio suavis*, or soft sell.

Lowes Dickinson was right, if we take him in a narrower sense than he may have intended. Science and imaginative *writing* are utterly incongruous, in English anyway – the French tradition is more permissive – and the effect of combining them is merely absurd. In science the imaginative element lies in the conception, and not at all in the language by which the conception is made explicit or is conveyed. (The 'language' might indeed use the symbolism of chemistry, mathematics, or electronic circuitry.) Clarity can be, *must* be achieved, and with a natural stylist like D'Arcy Thompson, grace. But a scientist's fingers, unlike an historian's, must never stray toward the diapason, and a falling cadence is allowed only to mark, and perhaps be the welcome evidence of, the end of a 'presentation'.

By the time of the New Philosophy, the competition or disputation between eloquence and wisdom, style and substance, medium and message had already been in progress for nearly 2,000 years, but as far as the New Philosophy was concerned, the Royal Society, with the formidable support of John Locke and Thomas Hobbes, may be thought to have settled the matter once and for all: scientific and philosophic writing were on no account to be made the subject of a literary spectacle and of exercises in the high rhetoric style.

This position has been threatened only during those two periods in which our native philosophic style (which is also a style of thinking) was obfuscated by influences from abroad. During the Gothic period of philosophic writing, which began before the middle of the nineteenth century and continued until the First

[1] Richards, *op. cit.* p. 18, n. 1.

World War, we were all oppressed and perhaps mildly stupefied by metaphysical profundities of German origin. But although those tuba notes from the depths of the Rhine filled us with thoughts of great solemnity and confusion, it was not as music, thank heavens, that we were expected to admire them. The style was not an object of admiration in itself. Today, though we are now much better armed against it, speculative metaphysics has given way to what might be called a *salon* philosophy as the chief exotic influence, and French writers enjoy the reverential attention that was at one time thought due to German. Style has now become an object of first importance, and what a style it is! For me it has a prancing, high-stepping quality, full of self-importance; elevated indeed, but in the balletic manner, and stopping from time to time in studied attitudes, as if awaiting an outburst of applause. It has had a deplorable influence on the quality of modern thought in philosophy and in the behavioural and 'human' sciences.

The style I am speaking of, like the one it superseded, is often marked by its lack of clarity, and hereabout we are apt to complain that it is sometimes very hard to follow. To say as much, however, may now be taken as a sign of eroded sensibilities. I could quote evidence of the beginnings of a whispering campaign against the virtues of clarity. A writer on structuralism in the *TLS* has recently suggested that thoughts which are confused and tortuous by reason of their profundity are most appropriately expressed in prose that is deliberately unclear. What a preposterously silly idea! I am reminded of an air-raid warden in wartime Oxford who, when bright moonlight seemed to be defeating the spirit of the blackout, exhorted us to wear dark glasses. He, however, was being funny on purpose.

I must not speak of obscurity as if it existed in just one species. A man may indeed write obscurely when he is struggling to resolve problems of great intrinsic difficulty. This was the obscurity of Kant, one of the greatest of all thinkers. There is no more moving or touching passage in his writings than that in which he confesses that he has no gift for lucid exposition, and expresses the hope that in due course others will help to make his intentions plain.[1]

In the eighteenth century obscurity was regarded as a disfigurement not merely of philosophic and scientific but also of theological prose. To conceal meaning (it was reasoned) is equally to conceal lack of meaning, so we don't know where we stand. George Campbell[2] (the Scottish philosopher and divine, not the poet) thought himself specially afflicted by mystical theology, and his interpretation of it will pass very well today. Mystical theology is a prose offering to the Almighty; and just as it is in the nature of a living sacrifice that it should be deprived of life before being offered up to the Godhead, so a prose offering must be deprived of – sense. This then is constitutive obscurity: that which appears to be nonsense for the simplest of all reasons, namely, that it is fact not sense.

But even in those enlightened days, the appeal of obscurity was clearly recognized. Of Dryden – even of Dryden – Johnson said that 'he delighted to tread upon the brink of meaning, where light and darkness begin to mingle'.[3] Don't we all, up to a point? We all recognize a voluptuary element in the higher forms of incomprehension and a sense of deprivation when matters which have hitherto been mysterious are now made clear.

The *rhetorical* use of obscurity is, however, a vice. It is often said – and it was said of Kant[4] – that the purpose of obscure or difficult writing is to create the illusion of profundity, and the accusation need not be thought an unjust one merely because it is trite. But in its more subtle usages, obscurity can be used to create the illusion of a deeply reasoned discourse. Suppose we read a text with a closely reasoned argument which is complex and hard to follow. We struggle with it, and as we go along we may say, 'I don't see how he makes that out', or 'I can see now what he's getting at', and in the end we shall probably get there, and either

[1] Immanuel Kant, *Critique of Pure Reason*, introduction to the second edition (1787).

[2] George Campbell, *The Philosophy of Rhetoric* (1776).

[3] Samuel Johnson, 'Dryden', in *The Lives of the Poets*, vol. 2 (1781).

[4] See Kant's preface to *The Metaphysic of Morals* (1797).

agree with what the author says or find reasons for taking a different view. But suppose there is no argument; suppose that the text is asseverative in manner, perhaps because analytical reasoning has been repudiated in favour of reasoning of some higher kind. If now the text is made hard to follow because of *non sequiturs*, digressions, paradoxes, impressive-sounding references to Gödel, Wittgenstein, and topology, 'in' jokes, trollopy metaphors, and a general determination to keep all vulgar sensibilities at bay, then again we shall have great difficulty in finding out what the author intends us to understand. We shall have to reason it out therefore, much as we reasoned out Latin unseens or a passage in some language we don't fully understand. In both texts some pretty strenuous reasoning may be interposed between the author's conceptions and our understanding of them, and it is strangely easy to forget that in one case the reasoning was the author's but in the other case our own. We have thus been the victims of a confidence trick.

Let me end this section with a declaration of my own. In all territories of thought which science or philosophy can lay claim to, including those upon which literature has also a proper claim, no one who has something original or important to say will willingly run the risk of being misunderstood; people who write obscurely are either unskilled in writing or up to mischief. The writers I am speaking of are, however, in a purely literary sense, extremely skilled.

IV

Let me now turn to a comparison between scientific and poetic notions of the truth, though only as far as it may help to recognize and define the literary syndrome in scientific or quasi-scientific thought.

When the word is used in a scientific context, *truth* means, of course, correspondence with reality. Something is true which is 'actually' true, is indeed the case. This is empirical truth – truth in the sense in which it is true to say that I am at this moment delivering the Romanes Lecture and not standing on my head on an ice floe in the North Atlantic; and you know that correspondence with reality in just this sense is the test that all scientific theories must be put to, no matter how lofty or how trivial they may be.

We must at once dismiss the idea that empirical or factual truth as scientists use it (or lawyers or historians) is an elementary or primitive notion of which everyone must have an intuitive or inborn understanding. On the contrary, it is very advanced, very grown up, something we learn to appreciate, not something that comes to us naturally. We must also, I think, dismiss the inductive interpretation of the way in which truth enters into scientific inquiry. In classical inductive theories of scientific method, plain factual truth is what scientific reasoning is supposed to begin with. We start (or else it is no use starting) with an exact apprehension of the facts of the case, with a reliable transcript of the evidence of the senses which inductive reasoning can thereupon compound into more general truths or natural laws. We are led into error (according to inductive theory) only when the facts we thought we could rely upon were wrongly apprehended. Error is due to an indistinctness of vision, a false reading of that Book of Nature in which the truth resides and can be got at if only we can retain or re-acquire the innocent, candid, childlike faculty of grasping what is in fact the case.

I share Karl Popper's view[1] that this conception of truth and error is utterly unrealistic. Scientific theories (I have said) begin as imaginative constructions. They begin, if you like, as stories, and the purpose of the critical or rectifying episode in scientific reasoning is precisely to find out whether or not these stories are stories about real life. Literal or empiric truthfulness is not therefore the starting point of scientific inquiry, but rather the direction in which scientific reasoning moves. If this is a fair statement, it follows that scientific and poetic or imaginative accounts of the world are not distinguishable in their origins. They start in

[1] Karl Popper, 'On the Sources of Knowledge and of Ignorance', reprinted in *Conjectures and Refutations* (London, 1963).

parallel, but diverge from one another at some later stage. We all tell stories, but the stories differ in the purposes we expect them to fulfil and in the kinds of evaluations to which they are exposed.

The divergence of poetic from factual truthfulness was not always taken for granted. For Sir Philip Sidney and his contemporaries it was something that had to be justified and reasoned out. 'Now for the poet,' says Sir Philip Sidney in a famous passage of his *Apology*, 'he nothing affirms and therefore never lieth. For, as I take it, to lie is to affirm that to be true which is false ... but the poet (as I said before) never affirmeth.' If, all other things being equal, the choice is between correspondence with and departure from reality, then the choice is for reality: a painting which professes to be a portrait must be a likeness; but, if the choice is between what things are and what they ought to be, 'considered in relation to use and learning', then the literal truth, what actually happened, is usually less *doctrinable* than things as they might have been. For the scientist (Sidney says 'historian', but in this context scientist will do) is in bondage to the particular, to that which was – the historian's 'bare *was*' is Sidney's phrase – and any precept or general statement compounded of these bare particulars can only have the force of a 'conjectured likelihood'. It will not have the force of a poetic truth.

The idea that a poetic truth is a revelation of the ideal, of what *ought to be*, is taken by Sidney from Aristotle. Sidney (and incidentally Bacon) construe *ought to be* in the moralistic or doctrinary sense. For Bacon[narrative poetry 'feigns acts more heroical' than anything which actually happened, and thus 'conduces not only to delight but also to magnanimity and morality'. Dramatic poetry may be 'a means of educating men's minds to virtue', and the purpose of what Bacon describes as the highest form of poetry, the parabolical, must obviously be to improve.

This is what Sidney understood by the concept of what *ought*

¹ Francis Bacon, *De augmentis scientiarum*, book 2, chap. 13.

to be,[1] but according to Professor Butcher's well-known analysis of the matter, it was not Aristotle's. The reason why Aristotle believed poetry to be 'a more philosophical and a higher thing than history' (and here, too, we may read 'science') is because it reveals what ought to be in the light of a true understanding of nature's intentions – not of nature's actions, for these are clumsy and imperfect. No, the poet discerns the purpose which nature is working, often most imperfectly, to fulfil. The poet is thus one up on nature (this was not Professor Butcher's expression) and is the spokesman of her unfulfilled designs.

Aristotle's conception enriches or replaces scientific truth by truth of a higher kind, that which represents the testimony of a deeper and more privileged insight – a truth so lofty that, if nature does not conform to it, why then, so much the worse for nature.

A second interpretation of poetic truth – the one I have just outlined is no longer professionally defended, which is not to say that it is no longer believed in – would claim for it that it represents truth not of a higher kind, but simply of a different kind, an alternative conception, or one of a set of alternatives, which enriches our understanding of the actual by making us move and think and orientate ourselves in 'a domain wider than the actual'. I believe this view is essentially a fair one, and it would be silly to squabble over matters of copyright to do with the usage of the word 'truth'. Nevertheless, great difficulties arise when it is allowed to infiltrate into science.

In this second conception of truth, a structure of imaginative thought – for example, a myth, especially if it appeals to magical agencies – will be judged true if it is all of a piece, hangs together, doesn't contradict itself, leaves no loose ends, and can cope with the unexpected. No single word in common speech describes this set of properties, but a narrative or theory or world picture or imaginative structure of any kind which answers to them is

[1] See S. H. Butcher, Aristotle's *Theory of Poetry and Fine Art*, 1st ed. (1894); see also Ingram Bywater, *Aristotle on the Art of Poetry* (Oxford, 1909), and D. S. Margoliouth, *The Poetics of Aristotle* (London, 1911), 'What ought to be' is so rendered by Butcher and Bywater; Margoliouth writes 'the ideal'.

said to 'make sense', to have the property of being *believable-in*. All scientific theories must make sense, of course, but in addition they are expected to conform to reality, to be empirically true. It is the relaxation of this condition, or the failure to enforce it, that opens up to us a world that is larger, more various, and perhaps more doctrinable than real life.

I spoke of myths. In his famous work on savage thought, C. Lévi-Strauss[1] dismisses the cosy, traditional belief that myths are primitive absurdities, are silly, innocent constructions that represent a merely rudimentary stage in the development of scientific thought. On the contrary, one can think of 'the rigorous precision of magical thought and ritual practices as an expression of the unconscious apprehension of the *truth of determinism*, the mode in which scientific phenomena exist'. Instead of contrasting magic and science, 'it is better to compare them as two parallel methods of acquiring knowledge', or as 'two scientific levels at which nature is accessible to scientific inquiry', both being 'equally valid'.

What Lévi-Strauss is telling us is that myths make sense, as conventional scientific theories make sense, and he does not feel that their failure to measure up to reality – to pass that extra examination which has to do with conformity to real life – disqualifies them from being described as 'scientific'. Some Siberian peoples, he tells us, 'believe that the touch of a woodpecker's beak will cure toothache', and for this and similar reasons,

it may be objected that science of this kind can scarcely be of much practical effect. The answer to this is that its main purpose is not a practical one. It meets intellectual requirements rather than or instead of satisfying needs ... The real question is not whether the touch of a woodpecker's beak does in fact cure toothache. It is rather whether there is a point of view from which a woodpecker's beak and a man's tooth can be seen as 'going together' (the use of this congruity for therapeutic

[1] C. Lévi-Strauss, *La pensée sauvage* (Paris, 1962), trans. as *The Savage Mind* (London, 1966).

c

purposes being only one of its possible uses), and whether some initial order can be introduced into the universe through these groupings.

This is a clear statement of his case, and I find it utterly unconvincing. Whose 'intellectual requirements' are being met, we may wonder, the savage's or the anthropologist's? By what extra criterion shall we be satisfied that the anthropologist himself is not creating a metamythology, a mythology about myths? And would not someone actually suffering from toothache incline toward a more pragmatic style of thought? The point is that making sense and being believable-in are necessary but not sufficient qualifications for a process of intellection to be called commonsensical or scientific. The world of myths is Blake's world, Beulah, 'a place where contrarieties are equally true',[1] a world where the opposite of truth is not falsehood, but another truth; not necessarily a rival truth, but the telling of a different story, the testimony of a different interpretation of the world. Another myth, another set of magical allegiances, may serve the same or an equivalent purpose. The evidence Lévi-Strauss brings forward to contest the commonplace view that myths are a kind of fumbling approximation to science – a first groping attempt to make sense out of the complexities of the world – is just that which seems to me to justify it. For myths are not really truths; at best they are truthlike structures, a part of the candidature for what might pass as true, but a candidature excused from public examination.

The insufficiency of merely making sense and conferring order is not always fully grasped by laymen. Freudian psychoanalytic theory is a mythology that answers pretty well to Lévi-Strauss's descriptions. It brings some kind of order into incoherence; it, too, hangs together, makes sense, leaves no loose ends, and is never (but never) at a loss for explanation. In a state of bewilderment it may therefore bring comfort and relief. But what about its therapeutic pretensions? The embarrassment of the woodpecker's beak

[1] Blake, *Milton*, book 2, pl. 30.

is now got out of most adroitly. For in the opinion of many advanced thinkers, it is rather – well, rather *common* to suppose that the purpose of Freudian psychotherapy is, in the conventional sense, to cure. Its purpose is rather to give its subject a new and deeper understanding of his own condition and of the nature of his relationship to his fellow men. A mythical structure will be built up around him which makes sense and is believable-in, regardless of whether or not it is true. Another such structure might do as much for him – or as little.

In existential psychiatry,[1] the idea of 'cure' is dismissed contemptuously and replaced by the idea of 'healing'. A madman, for example, is healed when a microcosm of thought and personal relationships is built up around him in which his behaviour is no longer 'mad', that is, incongruous, anti-social, alienated from the majority opinion. The concept of 'explanation' is replaced by that of *understanding*, the process of discernment that uncovers a scheme of thinking within which a madman's actions and opinions now make sense.

With Freudian or existential psychology, as with myths, the question hardly arises of rational agreement or disagreement: these are ugly, hectoring words. Rather it is a question of acquiescence, of being taken into the author's scheme of thinking – and to describe acquiescence as a process of being 'taken in' has exactly the right connotation of surrender on the one hand and on the other hand of magic or contrived illusion. For these well-intentioned people are *telling stories*, sometimes wonderfully imaginative stories and sometimes wonderfully well told, so perhaps we should exercise a grown-up indulgence. When children don't tell the truth, their mother doesn't summon them to her knees and call them flaming little liars. On the contrary, she says, 'You mustn't tell stories'; but although she wears a special kind of solemn face when she says so, she doesn't really

---

[1] My authorities here are mainly D. Cooper, *Psychiatry and Anti-Psychiatry* (London, 1967), and R. D. Laing, *The Self and Others* (London, 1961) and *The Divided Self* (London, 1960); and the occasional writings of both. See also M. Foucault, *Madness and Civilization*, English trans. (London, 1967), and J. Lacan, *Écrits* (Paris, 1966).

think that telling stories is wicked unless it actually leads to harm.

Unfortunately, the psychologies I have been talking about are highly mischievous, not so much because they do harm or fail to do good, but because they represent a style of thought that will impede the growth of our understanding of mental illness. Consider for a moment imbecility, a subject in which scientifically founded psychiatry has made some ground. Here is an imbecile child who when it was born seemed ordinary. What can be wrong? Did some immemorial foreknowledge of the intrinsic contradictions of living drive it back into the habitation of a voiceless inner world? Did its parents, by some involuntary withholding of compassion, fail to ratify the child's ontological awareness of its essential self? Or is it perhaps unable to metabolize phenylalanine? Does it have the right number of chromosomes? What about the concentration of triiodothyronine in its blood? Two quite different sets of questions, and the people who ask them belong to two quite different kinds of worlds, figuratively speaking, the salon and the laboratory. Cultural psychiatry (but here I exclude Freud himself) repudiates the idea of an organic cause of mental abnormality; repudiates, indeed, some of the very concepts in terms of which the notion is expressed. The scientist *wants* it to be true. If there existed in science and medicine an analogue of literary criticism, we should investigate not only what people have reason to believe in, but the kind of things they *want* to believe in, and the cultural history of how they have come to acquire two or more different habits of expectation which cannot be reconciled.

It follows from what I have said that Freudian and other quasi-scientific psychologies are getting away with a concept of truthfulness which belongs essentially to imaginative literature, that in which the opposite of a truth is not falsehood but (we are back in Beulah) another truth. I strongly suspect that the same may be true of the more literary forms of other behavioural sciences, but I have not studied them deeply enough to say so for sure.

V

My contention has been that science tends to expel literature, and literature science, from any territory to which they both have claims – particularly the areas of learning that relate to human behaviour in its widest sense.

The distinguishing marks of the literary syndrome in science are, if you have followed my argument, these. *First*, there is an open or implied claim to a higher insight than can be achieved by laboratory scientists or historians or philologists, or by philosophers of the traditional English kind, an insight which soars beyond the busy little world of test tubes and graphs and measuring instruments, or indeed of facts. *Second*, there is a combination of high imaginativeness with a relaxation of or a failure to enforce the critical process, so that the critical and inventive faculties no longer work together synergistically, but tend if anything to compete; and with this goes a whispering campaign against the importance attached to validation or justification and even, in extreme cases, now beyond remedy, against rational thought. *Third* (and this is what gives the syndrome a literary rather than a metaphysical character) is the *style* in which the high truths of the imagination are made known, a style which (among many other disfigurements) deliberately exploits the voluptuary and rhetorical uses of obscurity, a style which at first intrigues and dazzles, but in the end bewilders and disgusts.

One may well ask: if the forms of discourse that answer to this description are kept outside the reach of a critical apparatus; if in repudiating the ideas of proof or cure or any other scheme of validation they escape the sanctions that are enforced upon physicians or historians or laboratory scientists, what then is to stop them from expanding their influence and pretensions without limit? The answer is clear enough. They are not repudiated, but as fashion changes they will be forgotten, to be classified as a scientific curiosity or literary genre, as dead as the Philosophic Romances of the seventeenth century or the System Philosophies of the nineteenth century. This fate is the unhappiest that could befall them, because their practitioners want above all else to be in the swim, to be counted among the makers of cultivated opinion, rare spirits whose thought transcends the busy pre-occupations of common people. To be forgotten is the worst of their bad dreams.

There is an aberration of science or of the scientific style of thinking which has come to be known as 'scientism'. Roughly speaking, it stands for the belief that science knows or will soon know all the answers, and it has about it the corrupting smugness of any system of opinions which contains its own antidote to disbelief. I suppose my lecture has been about *poetism*, an aberration of imaginative literature about which (*mutatis mutandis*) one could say very much the same. It stands for the belief that imaginative insight and a mysteriously privileged sensibility can tell us all the answers that are truly worthy of being sought or being known, and its practitioners are rallied by the inane war cry that beauty is equivalent to truth.

For scientism, imaginative literature is best thought of as a branch of the entertainment industry; for poetism, scientists are engaged in merely parsing the Book of Nature, the inner meaning of which they are altogether unqualified to comprehend. Poetism is only a minor ailment of literature, but an ailment that literature is prone to through an excess of its own exuberant strength; in the time scale of literature, its outbreak seems to be a seasonal event. Scientism, for just the same reason, is latent in scientific thinking – a malady to which, because of our constitutions, we scientists are specially predisposed. Both views are about equally contemptible, so there is no need for us to take sides. No one need beat his breast and say, 'Now *I* am on the side of the poets', because poets are not really on that side and scientists not really on the other. I admit that it was mainly a love of science that prompted me to speak as I have done, but it could equally well have been a love of literature, and if it had been so I do not think that my lecture would have been so very different from that which you have just heard.

A Reply to Sir Peter Medawar
by John Holloway

If one is cast down – as I am not – by unfavourable circumstances, it might be wise to give up the profession of letters. Many of the best British Isles authors of today now live in Paris, America, or Ireland: the man of letters finds that his chosen field is silently emptying. In the university, some of his students will be intimating that reality has left him high and dry, and others will be enthusing over the great English classics not in the original but as television serials. If he looks into the future, he sees the landscape, the traditions, the *mores*, and even the human nature which have created his subject all being either transformed by science, or obliterated by its achievements, or terminated by famine. When he then finds that an eminent scientist has given a Romanes Lecture on 'Science and Literature', he may well begin to read in expectation of one more attack, this time from a new quarter. I choose my words carefully when I say that he will have to work to find it.

A hundred years ago Bishop Wilberforce attacked science; and at the hands of T. H. Huxley, a Romanes Lecturer perhaps even more distinguished than the present one, he fared extremely ill. His weapons were the weapons of superciliousness, and the man of letters will know better than to use these today. More likely he will feel quite the opposite. He will feel a great sense of inadequacy, partly because when Professor Medawar demands to know whether (to take one example) imbecility isn't perhaps a matter of triiodothyronine, the man of letters will sense his own ignorance on the one hand, and the great collaborative expertise of science on the other; and partly for a more attractive reason – more attractive, though sufficiently daunting all the same. This is, that 'Science and Literature' is really a gem of a lecture. It is years since one has read anything like it. Sir Peter knows all about triiodothyronine and the like, but there is also nothing that any man of letters can teach him about presenting a case and enchanting an audience. 'I hope I shall not be thought ungracious . . .' the lecture begins. This lecturer can never have been thought ungracious in his life.

Or again, 'Science and literature . . . where they might be expected to co-operate, they compete. I regret this very much [and] don't think it necessary . . . We all want to be friends.' This sort of graciousness is among the man of letters' weapons. Now he finds them turned against himself; and if he wants to take issue, must seem to start the unco-operativeness. Elsewhere, mankind's failure 'to justify science in any sense except the scientific' is described as 'a reproach to all scholars, scientists and humanists alike'. Just as the man of letters begins to wonder whether he ought to accept, along with scientists, a share of the blame – he finds it is coming his way altogether. 'I cannot even think of a name for the new discipline that might fill those empty spaces,' the lecturer proceeds, '. . . *perhaps the world I want is just criticism.*' Poor critics, it is them after all.

I have been writing rather as if the man of letters, when he studies this lecture, is going to think that he has to meet, or maybe to issue, some kind of challenge. Professor Medawar would almost certainly repudiate this. Another of the more gracious passages in his lecture is the closing sentence, where with beautiful lightness and discretion he intimates that he is no enemy but a lover of literature, and that what he has said is all in accord with that fact. Indeed, as I suggested earlier on, 'Science and Literature' is so good as a lecture that it is itself a piece of literature. But all the same, lovers sometimes need to be received with a little circumspection, and I think we have a case here.

It may be said that Professor Medawar puts science and literature very much on an even footing. 'My contention has been', he says in his concluding section, 'that science tends to expel literature, and literature science, from any territory to which they both lay claim.' Elsewhere we read of 'territories of thought . . . including those upon which literature has a proper claim', and of 'large territories of human belief and learning upon which both science . . . and literature have very important things to say'. Even though one hesitates over the idea that literature has important things to say about large areas of human *learning*, and may wonder in passing whether the sentence had (as it were) been composed for the sake of science, and literature was then slipped in for the sake of fairness, all the same, the impression of an open mind and an even-handed treatment is strong.

I find myself, though, with the difficult task of arguing that this is an illusion: that Professor Medawar, seeming and meaning also to hold the balance even, is putting his thumb down on it all the time. Perhaps he might castigate this way of putting the matter as among the 'trollopy metaphors' that he associates with what he calls – too loosely, I believe – '*the* literary style'; or elsewhere, 'imaginative writing'. But if it comes to that, 'trollopy metaphor' is itself a trifle *demi-mondaine*. In fact, for all his dislike of metaphors, this lecture is rather full of them, ('metaphysical profundities . . . *tuba-notes* from the depths of the Rhine'; 'the inane *war-cry* that Beauty is equivalent to Truth'); and sometimes, even though indirectly, they contribute, as with 'two quite different worlds, the *salon* and the laboratory', to one's sense that though the balance seems to stay even, this is not quite so. Particularly is this the case with one point in the argument which at first seems especially even in treatment. 'Scientific theories . . . begin, *if you like*, as stories . . . we all tell stories.' But on reflection, I find that I do not like; because later on, when the metaphor comes again, the balance goes down. The scientist's final story is one 'about real life'; but in the end the writer seems to 'tell stories' in another sense, and it seems that this has something to do with what a mother reproves her child for ('telling stories'); though, especially if the stories are 'wonderfully well told', she does so with 'grown-up indulgence'. That word 'grown-up' comes elsewhere: 'factual truth as scientists use it . . . is very advanced, very grown-up'. By now, one's sense of an evenly kept balance is less assured than at first.

These may seem very small points: but I am trying, little by little, to open out this beautifully sustained texture of argument and implication, and to show that Professor Medawar is much less a friend of literature than we think at first, or than (as I firmly believe) he thinks all the time. Another point of ingress might be what he says about English literature directly: because it is really quite a surprise to notice how unsympathetic this is for the most part. 'At a time when the writing of English had reached a peak of *adventurousness and effulgence.*' When is this? One cannot tell for certain, but it looks like the time of Shakespeare and Donne – and certainly it is followed, to the lecturer's relief, by 'the New Philosophers of the Seventeenth Century'. Between them and Blake, things are better: but when Professor Medawar notices an exception his tone is revealing:

> Johnson said that Dryden 'delighted to tread upon the brink of meaning . . .' Don't we all, up to a point? We all recognize a voluptuary element in the higher forms of incomprehension, and a sense of deprivation when matters that have hitherto been mysteries are now made clear.

But don't we also all recognize, in these words, a clear if unintentional hint of the indulgent 'grown-up' who is 'understanding' the child in others, or perhaps even in himself?

When Professor Medawar turns to the Romantic period, he is very bad; it is at this point that one of the two radical confusions in this lecture begins to show through the skilful surface. At the same time, one must admit that there has been a good deal of bad, programmatic writing on this period, pronouncing (like Professor Medawar) on that non-existent entity the 'official Romantic view', and so on, by critics and men of letters; and I am far more embarrassed for my colleagues who have lapsed into these first-year undergraduate errors, than ever Professor Medawar need be for his own imperfections. But all the same, it is just no good saying that the official Romantic view was that Reason and

Imagination are antithetical, when there is that celebrated passage, at the end of Wordsworth's *Prelude*, claiming that they are really two names for one thing. Nor can one blithely assert that the Romantics missed 'the synergism ... between imagination and reasoning, between the inventive and the critical faculties', and even level the charge at Coleridge by name, when Coleridge wrote:

> Imagination ... first put in action by the will and 'understanding', and *retained under their irremissive ... control*, reveals itself in ... a more than usual state of emotion *with more than usual order;* judgement ever awake and steady self-possession with enthusiasm and feeling profound or vehement.[1]

Wordsworth was equally emphatic in an equally celebrated passage:

> ... *habits of meditation* have ... prompted and regulated my feelings. For all [i.e. although] good poetry is the spontaneous overflow of powerful feelings: and though this be true, Poems to which any value can be attached were never produced on any variety of subjects but by a man who ... had *also thought long and deeply*.[2]

Even so, Wordsworth was not always able to satisfy Keats in these matters:

> ... it seems to me that if Wordsworth had *thought a little deeper* at that moment he would not have written [that] Poem at all ... it is ... not a *search after Truth*.[3]

Clearly enough, the emphasis is just the same as when Keats said, of one of his own poems, 'I assure you that when I wrote it it was a *regular stepping* of the Imagination *towards a Truth*' (Letter to Taylor, January 1818).

Professor Medawar reproves Arnold for the same blindness as

---

[1] *Biographica Literaria*, chap. 14.
[2] Preface to *Poems* (1800).
[3] *Letter to Bailey*, October 1817.

he wrongly attributes to 'the Romantics'. But when Arnold praised a poet in the highest terms he knew, it ran like this:

> the peculiar [i.e. uniquely excellent] characteristic of the poetry of Sophocles is . . . that it represents the highly developed human nature of [Sophocles'] age – human nature developed in a number of directions, politically, socially, religiously, morally developed – *in its completest and most harmonious development in all these directions*; while there is shed over this poetry the charm of that noble serenity which always accompanies true *insight*.[1]

Really, I do not think Professor Medawar would be satisfied if I were to sally into triiodothyronine territory, and boned up on it no better than this.

Yet the curious thing is that he probably need not have been drawn into these matters at all. Probably, he could have dealt with what he really wanted to deal with, and not taken on the Romantics or anything else of the kind. This is where the basic confusion comes out. Something which he calls the 'literary syndrome', for example, he describes like this:

> *First*, an open or implied claim to . . . an insight which soars beyond the busy little world of . . . facts; *Second*, . . . the critical and inventive faculties no longer work together . . . but tend if anything to compete . . .; *Third*, . . . a style . . . which . . . systematically exploits the voluptuary and rhetorical . . . which at first intrigues and dazzles, but in the end bewilders and disgusts.

Self-evidently, this has nothing to do with *Middlemarch*, or *King Lear*, or *Le Rouge et le Noir*, or *Anna Karenina*, or anything which anyone might set up as 'literature' over against 'science'. Nor was it part of Professor Medawar's purpose to do any such fatuous thing: he was not talking about any literary syndrome *tout court*, but about something that, wisely or unwisely, he called 'the literary syndrome *in science*'. Who may its practitioners be? Save

[1] 'The Modern Element in Literature'.

for Lévi-Strauss he names no names; and it is none of my business to replace this angelic reticence by something else. But phrases like 'salon philosophy', 'French writers', 'structuralism', 'existential psychiatry', and the like, make matters clear enough. What Professor Medawar really wanted to do was to attack certain forms of pseudo-science. And who am I to say him nay? Only, he seems to have disposed his forces hastily; and, noticing that science was the opposite of pseudo-science, and also in some sense the opposite of literature, to have taken it rather for granted that pseudo-science and literature were pretty much akin. When I think, however, of what he traces in his pseudo-scientific syndrome, what strikes me is how much it sounds not like literature, but like pseudo-literature. If he had arranged his lecture so as to contrast science with pseudo-science, and literature with pseudo-literature, and then to trace both a substantial contrast, and a substantial community, between science and literature in their true and valid forms – his difficulties might have been less. But now I am beginning to rewrite 'Science and Literature'. It is right to be generous to its author, but not quite so generous as that.

I suggested earlier that there were two radical confusions in 'Science and Literature'. This also was over-generous, because more can be found: like thinking that 'imaginative writing' is the same as letting one's fingers 'stray towards the diapason'. Does Professor Medawar suppose that the imagination which has helped to make him so eminent a scientist contributes nothing to his – in spite of occasional lapses – beautifully *non*-diapason style? Here is one more confirmation, though, of the fact that Professor Medawar, in spite of his admiration for the Royal Society and that celebrated enemy of the figurative, its historian Bishop Spratt, is himself, when it comes to metaphors, just a little of a 'voluptuary'; and it is a couple more of his *trouvailles* in this direction which help one to see the second radical confusion. This comes in what is perhaps the most interesting part of his lecture, in which he distinguishes scientific from literary or poetic or imaginative *truth*.

The first of these trollopy, voluptuary metaphors in which I sense a confusion is of course the story-telling metaphor itself; for it is clear that 'telling stories' can mean more things than one, and fairly clear that it clouds the issue in more ways than one. I think it would be right to say that the Scientist does not really *tell* the 'stories', the hypotheses, which in the course of his work he finds do not adequately 'fit the facts', though they at first gave some promise of doing so. It would be truer, I believe, to say that he considers, reviews, tries out, these subsequently rejected hypotheses. They are 'stories', I suppose, certainly in the sense of proving to be (like the child's fibs, possibly) untruths. The verified and established theory is surely not a 'story' in this sense at all. If it deserves to be called a 'story', that can only be because it is a story '*about real life*' (in Professor Medawar's phrase): a veridical narration. I myself incline to think that the metaphor lets one down here as well, since it seems to me that a scientific theory is not narrative in form at all. If I am right, then the 'telling stories' metaphor is no good for either the preliminary or the definitive stage of scientific theorizing.

The important confusion occurs, though, when Professor Medawar's story-telling idea is applied to literature. His claim is that science starts with a fiction (the unconfirmed hypothesis) and moves into truth – that is to say, something which is 'about real life', which will 'conform to reality'. Poetic or imaginative efforts, however, 'diverge' from this; as for conforming to reality, 'this condition is relaxed'. In other words, literature remains in all the stages of its production 'story-telling' in the original meaning of the term. Its function is not truth-telling in the empirical sense at all. And yet, it may be spoken of, along with science, as 'accounts of *the world*'! And yet science and literature have a common goal in that each has 'important things to say', in that *both* 'give an account of *ourselves* and investigate *our condition*'! The fact is that Professor Medawar seems to have found a metaphor (the word 'story') which can equally easily be made to mean either 'true statement' or 'false fiction', and has then treated it the way Mr Atlas used to treat iron bars.

The other troublesome metaphor is 'territory': 'there are large territories ... upon which both science and literature have very important things to say ... those [territories] upon which litera-ture has a proper claim ... any territory to which they both lay claim ...' This metaphor may not be open to objection in itself; but Professor Medawar is undoubtedly so when *he* uses it of literature, because of the distinction which he asserts between scientific, and literary or poetic, truth. Briefly, his claim is that while scientific truth must both 'make sense' (that is, satisfy a criterion of internal coherence) and also satisfy a criterion of 'correspondence with reality', with poetic truth, the latter con-dition is relaxed. The criterion of poetic truth is the coherence criterion alone. But if this is so, then not only can 'literature' (which equals 'poetry' in this discussion) not possibly have the same 'territory' as science; it cannot be said to have a territory at all. There is simply *no* independent reality, 'upon' which, or, more idiomatically, about which it can say something 'very important'. 'Upon' and 'about' are words in the language of correspondence-truth; and once we recognize this, we recognize that Professor Medawar's lecture is more heavily bedizened with metaphors even than it had seemed; for that literature 'opens up a *world* that is larger, more various than real life' proves to be one, and that it 'enriches our understanding of the actual by making us move and think and orientate ourselves in a *domain* wider than the actual' is another. World and domain cannot be understood here in any literal sense, or there would be something for the literary truth to correspond to. What transpires is simply this: Professor Medawar sometimes asserts, and sometimes implies, that literature is not 'about' anything at all; but at the same time constantly speaks as if it were either about the same reality as science (the reality which at one point, as I mentioned, he called 'real life'), or about another, wider, but perfectly genuine reality. Here is the second radical confusion.

But, confused or not, this is the most interesting point in the lecture, and that at which the discussion ought somehow to be pressed forward. I will try briefly, not indeed to conclude this discussion, but at any rate to open it; and I can do so by taking hints from several places in the lecture. First, one notices that Professor Medawar at one point says that established scientific theories are 'stories *about real life*'. To the man of letters, surely, it is transparently obvious that this is what he must say about many or most of the fictions of major literature. Curiously enough, he would surely want to add that 'about *real life*' (as opposed to 'about nature' or 'about reality') seemed to him if anything to be rather apter for the major literary work than for science. One need not pursue this in order to see that perhaps it points towards ways in which the correspondence criterion applies differently to literary-poetic truth from how it applies to scientific; but unquestionably applies all the same.

Second, there are two particularly interesting observations in this lecture, which come far apart, and which one is a little surprised to find left so. One is about literature and one about science, and they are significantly different yet curiously reminiscent of one another. Here they are:

1. [literary truth] enriches our understanding by making us move and think and orientate ourselves in a domain *wider than the actual* . . .

2. Scientific reasoning is . . . a dialogue . . . *between the possible and the actual*, between proposal and disposal . . . between *what might be true, and what is in fact the case*.

I find myself wondering whether Professor Medawar happened ever to set these remarks side by side in his mind. What seems to show up from his lecture, though, is that the 'possible', the 'what might be true', integral as it is to the scientist's thinking, is integral only to its earlier stages, and is 'disposed' of as the final stage is reached. But it seems, in contrast to this, that ideas of what is possible or what might be true, as against what is in fact the case, belong as much to the latest stage in the creation of the literary work – the final, finished version – as to any earlier stage; and that this is so not only in more ways than one, but in ways which require more analysis than they have received.

In saying this, I have first in mind what everyone knows: that a major literary fiction tells us 'about real life' indirectly, while directly it narrates not a historical case, but a 'convincing' possible case. About this, however, there is an important and much-neglected fact which needs analysis, though I am not able fully to supply it. It is that the 'case' (one recalls James's use of the word) narrated by a major fiction is not properly elucidated by any of the terms which are commonly shared between the scientific and the literary modes of thought. The cases of Anna and Vronsky, or Clarissa and Lovelace, or Birkin and Ursula – and so on – are quite certainly not 'average' cases; nor, equally certainly, are they 'probable' ones. In their different ways, they are all fairly, or very, exceptional cases; but at the same time, their interest is concealed by the bald use merely of the word 'possible'. Always, they bear on life more sharply than just to count as 'what *might* be true'; they are what might very well be so, or might only too easily be so, or something like that. They are extremes which somehow also have a quality of the classical; or seem revelatory of the nature of the normal, even as they deviate away from it to an extreme. (I suspect that here the man of letters might receive some kind of formal help from the applied mathematician; but on the only occasion when I sought such help, I was unable to explain my problem, and the matter had to lapse . . .)

There is another respect in which, to speak loosely, 'the actual' and 'the possible' come together in the literary work with a peculiar intimacy: and this also has been neglected. There is indeed a 'domain wider than the actual' in, say, the great novel or play, in that for much of the time, perhaps even all the time, the work asks us to sense how easily, how readily, it is open to more interpretations than one; how close to the way that the writer clearly wants us to see his action and his characters, are other ways of seeing them, which he probably wants us to reject – though if we reject these without coming near to accepting them first, his efforts are as good as wasted. Surely, also, it is at this point that one sees a reason why 'about real life' fits literature better even than it does science. This quality of things which one can express (though only loosely) by saying that possibility, risk, and

D

potentiality nudge right up against actuality at every point, is present and potent in literature because so it is also in the common affairs of men. Often, of course, ordinary life admits of such a clear Yes-or-No as I suppose the scientist has when he feels able firmly to eliminate a hypothesis; but one knows the terrible destructiveness of those who think it does so all the time.

Does this mean that Professor Medawar is a terribly destructive man? I am afraid that, like many other real-life questions, this does not quite admit of a clear yes or no. Certainly, that ordinary life is often of that kind means that one cannot always speak quite plainly and unequivocally about it; and – in intention if not always in performance – Professor Medawar is at his least tolerant over obscurity. In his view, unless one is (like Kant) struggling with intrinsically *difficult problems,* to write obscurely is either to lack skill, or to be 'up to mischief'. 'No one who has something . . . important to say will willingly run the risk of its being misunderstood.' In reply, one can only say, it depends what you think real life is like; and Professor Medawar seems not to have improved on George Chapman:

> Obscurities in affection of words and undigested conceits, is pedanticall and childish; *but where it shroudeth it selfe in the heart of [its] subject . . . with that darkness will I labour to be shrouded.*

This would never do for Professor Medawar. Of 'the concept of truthfulness which belongs essentially to imaginative literature, that in which the opposite of a truth is not a falsehood but . . . another truth', he says, with a touch of the scorn which one cannot but detect not infrequently when literature and its values are at issue – '*We are back in Beulah*'. But I do not believe that 'real life' is Beulah at all.

A Rejoinder

Twenty years ago, when I first applied for a visa to enter America, a consular official asked me if it was in any sense my ambition to overthrow the Constitution of the United States. I told him that I would certainly not overthrow it on purpose and could only hope that I wouldn't do so by mistake. He didn't seem to think this very funny, and I suppose Mr John Holloway won't either if I answer him in the same spirit; for he in his turn suspects me of scheming to overthrow literature, and asks himself whether I may not perhaps be a 'terribly destructive man'. I am sorry he should have misunderstood me so far as to entertain such horrid suspicions, and do not quite know why he thought it right to adopt such a wary and defensive attitude throughout. I am not attacking him or anything he stands for.

Holloway says that my attack was 'really' directed against pseudo-science and pseudo-literature. This is true in a sense, but in a sense that is needlessly inexact. Science and literature can be 'pseudo' in a great many different ways: science through fraud or self-deception, faulty reasoning, or any one of a number of methodological omissions or mistakes; literature through insincerity, inhumanity, triviality, hollowness, and the rest. In my Romanes Lecture I was speaking about a special kind of corruption of science ('poetism') that is characteristically literary in origin, and, by implication, about a special kind of pseudo-ness in literature and the humane arts generally ('scientism') that can be traced back to an origin in science. I expressed my abhorrence of them both. I was not trying to be gracious and conciliatory when I said, in its last sentences, that my lecture would not have been so very different if it had been inspired by love of literature rather than by love of science; but a defence of the humane arts against mistaken, misapplied or contextually inappropriate scientific notions would have been a much triter theme, and one that I am not specially well qualified to develop. And why should I play an away match when I don't think of myself as playing a match at all?

Holloway has cruelly exposed one or two indiscretions or errors of taste in my lecture. ('Trollopy metaphors', for example. I had doubts when I wrote it. Why *does* one let that kind of thing go by?) He too has lapses of taste, and misquotes me twice, on one occasion seriously. I did not say that 'imaginative writing is the same as letting one's fingers stray towards the diapason', nor would I express myself in such a clumsy way. What I said was, 'a scientist's fingers ... must never stray towards the diapason' – and indeed they must not. Nevertheless there are quite a number of things in my lecture which I now wish I had put otherwise, and Holloway is too good a critic not to feel the same about his reply. Let us turn to more serious matters.

First, the Romantics. Did Coleridge really grasp the idea of a synergism between reason and imagination; between the inventive and critical faculties? I do not wish to sound arrogant, but on this technical and rather specialized subject I am probably the better informed and better read. Holloway makes a bad case for Coleridge, if his quotation is the best evidence he can summon up. My discussion of the matter in my book, *The Art of the Soluble* (1967), shows how very much nearer he actually got. Holloway's quotations show, not that his chosen spokesman understood the gist of the hypothetico-deductive method, but that he himself does not. So far as I know the first writer in English to see scientific method as a dialogue between imaginative proposal and critical disposal, the two together doing what neither could do apart, was William Whewell in *The Philosophy of the Inductive Sciences* (1840). Coleridge had some important elements of that conception, but failed to grasp it as a whole.

I called this a tragedy of cultural history for two reasons: (*a*) Coleridge *could* have got there; he alone had everything it takes – the depth of learning, the intellectual penetration, the philosophic and scientific tastes; my admiration for him is such that I would stretch the evidence almost beyond reason to give him credit for a prescience which he did not in fact possess. (*b*) Whewell's scientific methodology was overshadowed for nearly 100 years by Mill's, and for reasons that deserve a special investigation by students of the history of ideas – because so much of the modern rivalry or sense of tension between science and literature has grown out of the idea that 'scientific method' is a precise formulary of intellectual behaviour, a calculus of discovery that supersedes imaginative insight in scientific thought, just as modern technology supersedes ineffectual homely skills.

Holloway derides my speaking of an 'official' Romantic opinion, missing (but this must be my fault) the ironic juxtaposition of the Romantic with the official, i.e. Royal Society, point of view. It is now seventy years since Professor Henry Beers told us that the English Romantic movement had 'no leader, no programme, no organ, no theory of art, and very little coherence'. No official opinion, then: the point is taken, though one wonders why it should be made.

Yet a belief in the antithesis between Reason and Imagination was a sort of colour-wash under almost the whole of English romantic thought. For Blake on imagination versus critical reasoning, refer in particular to his annotations of George Berkeley's *Siris*, or summon up from *Jerusalem* the Spectre of Rational Power that cast a mildew over Albion; for Blake the imaginative faculty, unlike reasoning, is inborn, innate, a manifestation of divinity in man; anyone who thinks otherwise is a fool or a knave.

The distinction Shelley draws, in his *Defence of Poetry*, between the proper domains of reason and imagination gains special force from his having told Elizabeth Hitchener ten years beforehand that he had 'rejected all fancy, all imagination . . . I am now an undivided votary of reason'. A man who changes sides must be specially aware that there are sides to change. The *Defence* was, of course, a defence against Peacock's *Four Ages*, but on this point they did not disagree. It is implicit in Peacock's argument that reason and imagination compete with each other; much of Shelley's defence is against the imputation that reason was gaining the upper hand. In his Introduction to the *Encyclopaedia Metropolitana*, Coleridge saw the poetic impulsion as 'a mighty, inward power, a feeling *quod nequeo monstrare, et sentio tantum*', and in Chapter XIII of *Biographia Literaria* there is a famous passage in which exercise of the 'primary imagination' is seen as a microcosmic rehearsal of the primordial act of creation itself, 'a repetition in the finite mind of the eternal act of creation in the infinite *I AM*'.

I am not quite sure what Holloway intends us to infer from Arnold's passage on Sophocles; the passages I myself had in mind were from *The Function of Criticism at the Present Time*, and here he will be found to express just the opinions I attribute to him.

To go back to synergism. Can it be that Holloway takes 'synergism' to mean no more than that a poet or other artist will scrutinize his own work very critically before he is satisfied with the result? I think not, because the synergism to which I was referring represented (so I said) 'the most important methodological discovery of modern thought'. For me the most interesting and exciting of all intellectual problems is *how* the imagination is harnessed for the performance of scientific work, so that the steam, instead of blowing off in picturesque clouds and rattling the lid of the kettle, is now made to turn a wheel. Literature and the fine arts have a cognate problem, but its solution, whatever it may be (and there must surely be more than one), is not the same.

It is towards a solution of that other great problem that Holloway, in his later paragraphs, is reaching out. I do not disagree with what he says and do not think that Aristotle would have done so either; my only complaint is that the context of his argument should be an attack on my own opinions. If he had said that my Romanes Lecture was not fully worked out and offers no definitive solution of any of the problems with which it deals, I should have agreed with him completely. I am groping: so is he. But it really is perverse of him to construe the lecture as a covert attack on literature in general. Of Poetic Truth I said that people might claim for it that it represents truth, not of a higher kind, but simply of a different kind; an alternative conception ... which enriches our understanding of the actual by making us move and think and orientate ourselves in 'a domain wider than the actual'. I believe this view is essentially a fair one ... [but] great difficulties arise *when it is allowed to infiltrate into science.*

Elsewhere I spoke of a literary syndrome *in scientific or quasi-scientific thought,* and quite early in my lecture I undertook to illustrate its bad influence on the behavioural *sciences.* Surely my intentions were clear enough?

Again, Holloway is very scornful when he attributes to me the opinion (I am quoting him now) that

science and literature have a common goal in that each has 'important things to say', in that *both* 'give an account of *ourselves* and investigate *our condition*'!

In fact, I said something quite different: that both science and literature have very important things to say about the behavioural sciences, and that

these subjects lie within the compass of literature in so far as they have to do with human hopes, fears, beliefs and motives; with the attempt to give an account of ourselves and investigate our condition ...

Holloway and I are very far apart here: I simply don't know what is preying on his mind.

Holloway ends his critique with George Chapman on obscurity. I do not know the context, so cannot say whether I have improved on him or not. As a medical scientist I have always taken Thomas Browne as chief spokesman for the voluptuary element in incomprehension ('I love to lose myself in a mystery ...' etc.) and am quite sure I cannot improve on what Coleridge had to say on the matter (*Aphorisms on Spiritual Religion,* VIII).

But in all this disputatious talk about Reason and Imagination and what reality really consists of, is there not a danger that we may overlook what George John Romanes had to say himself?

Let me make amends for having declined, in my lecture, to quote his poetry. Here is a fragment of Romanes's memorial poem to Darwin. We are to imagine Romanes in a paroxysm of grief at the sheer unreasonableness of Darwin's being taken away from us.

> '*The struggle cease,*' he counsels himself,
>> *And when the calm of Reason comes to thee,*
>> *Behold in quietness of sorrow peace.*
>> *By such clear light e'en in thine anguish see*
>> *That Nature, like thyself, is rational;*
>> *And let that sight to thee such sweetness bring*
>> *As all that now is left of sweetness shall:*
>> *So let thy voice in tune with Nature sing,*
>>> *And in the ravings of thy grief be not*
>>> *Upon her lighted face thyself a blot.*

This poem was specially selected for us by a distinguished man of letters and a former President of Magdalen College, T. Herbert Warren; *why*, God only wot.

Further Comments on
Psychoanalysis

In my Romanes Lecture on Science and Literature I implied that a psychoanalytical explanation-structure answered pretty closely to Lévi-Strauss's description of a myth. By this I meant that a psychoanalytical interpretation weaves around the patient a well-tailored personal myth within the plot of which the subject's thoughts and behaviour seem only natural, and, indeed, only what is to be expected.

I must begin by making it clear that my criticism of psychoanalysis is not to be construed as a criticism of psychiatry or psychological medicine as a whole. People nowadays tend to use 'psychoanalysis' to stand for all forms of psychotherapy, much as 'Hoover' is used as a generic name for all vacuum cleaners and 'Vaseline' for all ointments of a similar kind. By psychoanalysis I understand that special pedigree of psychological doctrine and treatment which can be traced back, directly or indirectly, to the writings and work of Sigmund Freud. The position of psychological medicine today is in some ways analogous to that of physical or conventional medicine in the middle of the nineteenth century. The physician of 120 years ago was confronted by all manner of medical distress. He studied and tried to cure his patients with great human sympathy and understanding and with highly developed clinical skills, by which I mean that he had developed to a specially high degree that form of heightened sensibility which made it possible for him to read a meaning into tiny clinical signals which a layman or a beginner would have passed over or misunderstood. The physician's relationship to his patient was a very personal one, as if healing were not so much a matter of applying treatment to a 'case' as a collaboration between the physician's guidance and his patient's willingness to respond to it. But – there was so little he could do! The microbial theory of infectious disease had not been formulated, viruses were not recognized, hormones were unheard of, vitamins undefined, physiology was rudimentary and biochemistry almost non-existent.

The psychiatry of today is in a rather similar position, because we are still so very ignorant of the mind. But the best of its practitioners are people of great skill and understanding and apparently inexhaustible patience; people whose humanity reveals itself just as much in the way they recognize their limitations as in their satisfaction when a patient gets better in their care.

I am emphasizing this point to make it clear that to express dissatisfaction with psychoanalysis is not to disparage psychological medicine as a whole.

One of my critics accused me of saying or implying that he, a psychoanalyst, would attempt to treat by psychiatric means the symptoms of a brain tumour or of Huntington's Chorea.

Of course I don't think a psychoanalyst would knowingly attempt to treat a brain tumour or a victim of Huntington's Chorea by psychoanalytic methods, but he may not realize the degree to which he is being wise after the event. Being a sensible man he naturally repudiates the idea of treating those psychological ailments of which physical causes are, in general terms, already known. But psychoanalysts do treat and speculate upon the origins of schizophrenic conditions and manic-depressive psychoses. *These* are the test cases: what are we to make of *them*?

Are 'mental illnesses' of mental or physical origin? To answer this question I shall begin with what may appear to be a digression. As recently as thirty years ago, many geneticists were still worried and confused by the problem of assessing, in precise terms, the relative contributions of nature and nurture – of heredity and environment or upbringing – to the overt ('phenotypic') differences between our mental and physical constitutions and capabilities. Both nature and nurture exercise an influence, of course; but L. T. Hogben and J. B. S. Haldane were the first to make it publicly clear that there is no *general* solution of the problem of estimating the size of the contribution made by each. The reason is that the size of the contribution made by nature is itself a function of nurture. (I use the word 'function' in its mathematical sense.)[1] If someone constitutionally lacks the ability to synthesize an essential dietary substance, say X, then the contribution made by heredity to the difference between himself and his fellow men will depend on the environment in which they live. If X is abundant in the food he normally has access to, his inborn disability will put him at no disadvantage and may not be recognized at all; but if X is in short supply or lacking, then he will become ill or die. The same reasoning applies to other, much more complicated examples. If people live a simple pastoral life that makes little demand on their resourcefulness and ingenuity, inherited differences of intellectual capability may not make much difference to their behaviour; but it is far otherwise if they live a difficult and intellectually demanding life. How often has it not been said that the stress of modern living raises the threshold of competence below which people can no longer keep up or make the grade? This is not to deny that some differences between us are for all practical purposes wholly genetic, wholly inborn. A person's blood group is described as 'inborn' not just because it is specified by his genetic make-up, but because (with certain rare and known exceptions) there is no environment capable of supporting life in which that specification will not be carried out. Most differences between us are determined both by nature and by nurture, and their contributions are not fixed, but vary in dependence on each other.[2]

With this analogy in mind, let me now turn to psychological disorders, which – to beg no questions – I shall define as conditions which cause a person to seek, or need, or be directed towards the

---

[1] In mathematics, x is a function of y when the value of x varies in dependence on the value of y.

[2] To speak (as I do here and below) of the causes of *differences* between human beings sounds clumsy and takes some getting used to; but there seems to be no avoiding it if one is to be precise and at the same time avoid a formal symbolic treatment.

care of a psychiatrist. Here, too, as a first approximation, it will be reasonable to assume that both 'mental' and 'organic' states or agencies contribute to the difference between the psychiatrist's patient and his fellow men; but here, too, we should be very cautious in our attempts to assign precise values to the contributions made by each. It seems natural to repudiate the idea of psychiatric treatment of brain tumours, because they seem so obviously organic in origin; but even in this extreme case we mustn't be too sure. Many of us now believe that there exists a natural defensive mechanism against tumours which is of essentially the same kind as that which prohibits the transplantation of tissues from one individual to another. If these natural defences are indeed immunological in nature, they are open to influences of a kind that common sense will classify as mental, or anyhow behavioural, e.g. to prolonged frustration, unhappiness, distress, or indifference to living. (The psychosomatic element in tuberculosis is specially relevant here, because the natural defence against tuberculosis depends on immunological mechanisms of a very similar kind.)

To go now to the other extreme: the psychoanalytic critic I referred to above thinks it probable that 'neurosis is the result of faulty early conditioning' rather than of brain disease or an inborn error of metabolism. No doubt; but does he not also think that constitutional or organic influences may raise or lower the susceptibility of his patients to these disturbing influences? Of course he does – and so did Freud. It is normally a mistake, I suggest, to trace any psychological disorder to wholly mental or wholly organic causes. Both contribute, though sometimes to very unequal degrees, and the contribution made by one will be a function of the contribution made by the other.

It is, nevertheless, very understandable that psychiatrists should approach their patients with two rather different kinds of etiological purpose and interest in mind. Psychiatrist A will say, 'My interest lies in trying to see how a certain pattern of upbringing, environment, habits of life and human relationships may predispose people of certain constitutions to psychological disorders.'

Psychiatrist B will say, 'Now *my* interest lies in trying to identify those elements of heredity and organic constitution which make a man specially likely to contract a certain psychological disorder if he is influenced by the environment and his fellow men in certain ways.' Both attitudes seem very reasonable, and over much of the territory that belongs to them the two psychiatrists will not compete. But – and now I come to my main point – in the context of those serious psychological disorders that are still disputed territory, the methodology implicit in the attitude of Psychiatrist B is very much the more powerful.

The reason is this. A physical abnormality can be the subject of diagnosis, and therefore, in principle, of treatment or avoiding action, *before* it can contribute to a psychological disturbance. The recognition early in life of a certain physical abnormality (say, the chromosomal constitution XYY) defines *a priori* a category of men who are at special risk; and our foreknowledge of that risk can be made the basis of a rational system of avoidance. The physical disability represents a parameter of the situation, where upbringing and environment are variables which can be varied within certain limits at our discretion. A difficult enterprise, to be sure; but not so difficult and much more realistic than, say, to abolish all family life, as one 'existential psychiatrist' is alleged to have recommended, because some families create an environment conducive to mental disorder. With certain forms of low-grade mental deficiency, this programme is now adopted as a matter of routine. When tests carried out on a baby's urine suggest that it cannot metabolize the amino-acid phenylalanine, its diet can be altered in such a way as to prevent or reduce the severity of what might otherwise be irremediable damage to the brain. I hope and expect that cognate solutions will one day be found for the major psychoses. No matter what other factors may have influenced him, there is something organically wrong with a manic-depressive patient, and it is essential to find out what it is, preferably before he becomes gravely ill.

This completes my attempt to explain why I think that the categorical distinction between brain disease and mental illness as between 'Nature' and 'Nurture' is a fundamentally unsound one – the remnant of an effete dualism, a still further perpetuation of what Ryle called the legend of Two Worlds.

I now turn to psychoanalysis itself, taken in the sense I gave it in an earlier paragraph. I shall not attempt a systematic treatment, but shall merely draw attention to a few of its more serious methodological, doctrinal and practical defects.

The property that gives psychoanalysis the character of a mythology is its combination of conceptual barrenness with an enormous facility in explanation. To criticize a theory because it explains everything it is called on to explain sounds paradoxical, but anyone who thinks so should consult the discussion by Karl R. Popper in *Conjectures and Refutations* (1963), particularly the passages (pp. 34–9) that make mention of psychoanalysis itself. Let me illustrate the point by a number of passages chosen from the authors' summaries of their own contributions to the 23rd International Psychoanalytical Congress held in Stockholm in 1963. I choose the Proceedings of a Congress rather than the work of a single author so as to get a cross-section of psychoanalytic thought.

> Character-traits are formed as precipitates of mental processes. They originate in innate properties; they come into existence in the mutual interplay of ego, id, super-ego and ego-ideal, under the influence of object-relations and environment.

> When an individual strikes out at his wife, his child, his acquaintances or even complete strangers, we may well suspect that a gross failure in Ego-functioning has occurred. Its restraining control has been partially eluded.

Of a 'cyclothymic' patient in the fifth and sixth years of psycho-analytic treatment:

> ... the delusion of having black and frightening eyes took the centre of the analytic stage following the resolution of some of the patient's oral-sadistic conflicts. It proved to be a symptom of voyeuristic tendencies in a split-off masculine infantile part of

the self and yielded slowly to reintegration of this part, passing through phases of staring, looking at and admiring the beauty of women.

On the etiology of anti-Semitism:

> The Oedipus complex is acted out and experienced by the anti-Semite as a narcissistic injury, and he projects this injury upon the Jew who is made to play the role of the father ... His choice of the Jew is determined by the fact that the Jew is in the unique position of representing at the same time the all-powerful father and the father castrated ...

On the role of snakes in the dreams and fantasies of a sufferer from ulcerative colitis:[1]

> The snake represented the powerful and dangerous (strangling), poisonous (impregnating) penis of his father and his own (in its anal-sadistic aspects). At the same time, it represented the destructive, devouring vagina ... The snake also represented the patient himself in both aspects as the male and female and served as a substitute for people of both sexes. On the oral and anal levels the snake represented the patient as a digesting (pregnant) gut with a devouring mouth and expelling anus ...

I have not chosen these examples to poke fun at them, ridiculous though I believe them to be, but simply to illustrate the olympian glibness of psychoanalytic thought. The contributors to this Congress were concerned with homosexuality, anti-Semitism, depression, and manic and schizoid tendencies; with *difficult* problems, then – problems far less easy to grapple with or make sense of than anything that confronts us in the laboratory. But where shall we find the evidence of hesitancy or bewilderment, the avowals of sheer ignorance, the sense of groping and incompleteness that is commonplace in an international congress of, say, physiologists or biochemists? A lava-flow of *ad hoc* explanation

---

[1] A disease of the kind psychoanalysts would be well advised not to meddle with.

pours over and around all difficulties, leaving only a few smoothly rounded prominences to mark where they might have lain. Surely the application of psychoanalytic methods in a completely alien culture might give even the most sanguine practitioner reason to pause? Not a bit of it. We have the word of two of the contributors to the Congress that 'the usual technique and theory of psychoanalysis were found to be applicable to obtain an understanding of the inner life' of the Dogon peoples in Mali:

> A twenty-four-year-old Dogon man, who at the beginning had met the white stranger with profound distrust, was led to change his views with surprising speed.
> After first having built a subsidiary transference and involved a younger colleague in the analysis, he turned from the animate object to the inanimate (playing with sticks) and from this to tactile gestures ... Finally he 'regressed' to somatic forms of expression in that he continued the analytic exchange by urinating ...

The examples I have chosen above, and the psychoanalytic autopsies I shall mention later, illustrate another important methodological defect of psychoanalytic theory. If an explanation or interpretation of a phenomenon or state of affairs is to be fully satisfying and actable-on, it must have a special, not merely a general relevance to the problem under investigation. It must be rather specially an explanation of whatever it is we want to explain, and not also an explanation of a great many other, perhaps irrelevant things as well.

For example: if a patient cannot retain salt in his body, it is not good enough (though it will probably not be wrong) to say that his endocrine system is in disorder, because such an explanation would cover a multitude of other abnormalities besides. The explanation may well be that the patient is no longer producing aldosterone, a specific hormone of the cortex of the adrenal gland, and if that is so he can probably be cured. Again, it will not do to say that muscular contraction is a transformation of energy derived originally from the sun. This is a weak explanation; it is too far removed in the pedigree of causes; we are more interested in the causal parentage of the phenomenon than in its causal ancestry. Strong explanations have a quality of *special* relevance, of logical immediacy: and this is a quality they must have if they are to be tested and shown to be acceptable for the time being or, as the case may be, unsound. Psychoanalytic explanations are invariably weak explanations in just this sense.

'Validation of psychoanalytic theory is a difficult business', my psychoanalytic disputant said, though he betrayed no logical understanding of why it should be so; and by implication he suggested that, instead of criticizing it destructively, I should help find means of testing whether or not it is true. Alas – except in one respect, which I shall deal with in a moment – the methodological obstacles are insuperable. Indeed, psychoanalysis has now achieved a complete intellectual closure: it explains even why some people disbelieve in it. But this accomplishment is self-defeating, for in explaining why some people do not believe in it, it has deprived itself of the power to explain why other people do. The ideas of psychoanalysis cannot both be an object of critical scrutiny and at the same time provide the conceptual background of the method by which that scrutiny is carried out.

It is for this reason that the notion of *cure* is methodologically so important. It provides the only independent criterion by which the acceptability of psychoanalytic notions can be judged. This is why cure is such an embarrassment for 'cultural' psychiatry in general. No wonder its practitioners try to talk us out of it,[1] no wonder they prefer to see themselves as the agents of some altogether more genteel ambition, e.g. to give the patient a new insight through a new deep, inner understanding of himself. But let us not be put off. Some people get better *under* psychoanalytic

---

[1] 'Curing is so ambiguous a term', says Dr David Cooper in *Psychiatry and Anti-Psychiatry*; 'one may cure bacon, hides, rubber, or patients. Curing usually implies the chemical treatment of raw materials so that they may taste better, be more useful, or last longer. Curing is essentially a mechanistic perversion of medical ideals that is quite opposite in many ways to the authentic tradition of healing.' Somewhat similar views are to be found in the writings of R. D. Laing, Michel Foucault, and J. Lacan.

E

treatment, of course; but do they get better as a specific consequence of psychoanalysis as such? I cannot condense the answer I gave in *The Art of the Soluble*, and therefore reproduce it here:

A young man full of anxieties and worries may seek treatment from a psychoanalyst, and after eighteen months' or two years' treatment may find himself much improved. Was psychoanalytic treatment responsible for the cure? One cannot give a confident answer unless one has reasonable grounds for thinking:

(*a*) that the patient would not have got better anyway;

(*b*) that a treatment based on quite different or even incompatible theoretical principles, e.g. the theories of a rival school of psychotherapists, would not have been equally effective; and

(*c*) that the cure was not a by-product of the treatment. The assurance of a regular sympathetic hearing, the feeling that somebody is taking his condition seriously, the discovery that others are in the same predicament, the comfort of learning that his condition is explicable (which does not depend on the explanation's being the right one) – these factors are common to most forms of psychological treatment, and the good they do must not be credited to any one of them in particular. At present there is no convincing evidence that psychoanalytic treatment as such is efficacious, and unless strenuous efforts are made to seek it the entire scheme of treatment will degenerate into a therapeutic pastime for an age of leisure.

The lack of good evidence of the specific therapeutic effectiveness of psychoanalysis is one of the reasons why it has not been received into the general body of medical practice. A layman might be inclined to say that we should give it time, for doctors are conservative people and ideas so new take ages to sink in. But it is only on a literary time scale that Freudian ideas are new. By the standards of current medical practice they have an almost antiquarian flavour. Many of Freud's principles were formulated before the recognition of inborn errors of metabolism, before the chromosomal theory of inheritance, before even the rediscovery of Mendel's laws. Hormones were unheard of when Freud began to propound his doctrines, and the mechanism of the nervous impulse, of which we now have a pretty complete understanding, was quite unknown.

Nevertheless, psychoanalysts are wont to say that Freud's work carried conviction because it was so firmly grounded on basic biological principles.

I am therefore sorry to have to express the professional opinion that many of the germinal ideas of psychoanalysis are profoundly unbiological, among them the 'death-wish', the underlying assumption of an extreme fragility of the mind, the systematic depreciation of the genetic contribution to human diversity, and the interpretation of dreams as 'one member of a class of *abnormal* psychical phenomena'.

I said earlier that the mythological status of psychoanalytic theory revealed itself in its combination of unbridled explanatory facility with conceptual barrenness, a property to which I have not yet referred. Ever since Freud's factually erroneous analysis of Leonardo, psychoanalysts have tried their hand at 'interpreting' the life and work of men of genius, and many of the great figures of history have been disinterred and brought to the post-mortem slab. The fiasco of Darwin's retrospective psychoanalysis has already been held up to ridicule.[1] But, Darwin apart, how can we not marvel at the way in which the whole exuberant variety of human genius can be explained by the manipulation of a handful of germinal ideas – the Oedipus complex, the puzzlement of discovering that not everyone has a penis, a few unspecified sado-masochistic reveries, and so on: surely we need a more powerful armoury than this? Evidently we do, for these analyses always stop short of explaining why genius took the specific form that interests us. Freud does not profess to tell us why Leonardo became an artist. 'Just here our capacities fail us', he says, with a modesty not found in the writings of his successors; but it is hard not to feel let down.

[1] In *The Art of the Soluble* (London, 1967), pp. 61-7.

A critique of psychoanalysis is, in the outcome, never much more than a skirmish, because (as I tried to explain) its doctrines are so cunningly insulated from the salutary rigours of disbelief. It is nevertheless customary to end any such critique with a spaciously worded acknowledgement of our indebtedness to Freud himself. We recognize his enlargement of the sensibilities of physicians, his having opened up a new area of human speculation, his freeing us from the confinements of prudery and self-righteousness, etc. There is some truth in all of this. There is some truth in psychoanalysis too, as there was in Mesmerism and in phenology (e.g. the concept of localization of function in the brain). But, considered in its entirety, psychoanalysis won't do. It is an end-product, moreover, like a dinosaur or a zeppelin; no better theory can ever be erected on its ruins, which will remain for ever one of the saddest and strangest of all landmarks in the history of twentieth-century thought.

The Genetic Improvement of Man

I make no apology for paying my respects to Macfarlane Burnet in the form of a philosophic dissertation on the genetic improvement of man. Burnet is a virologist, an epidemiologist, an immunologist and other things besides, but above all he is a biologist, a leader of biological thought. Because genetics, broadly conceived, is the central discipline of biology, and because evolution is its unifying doctrine, it follows that all biologists must be (or should be) interested in the notion of the genetic improvement of man. So having given some thought to the matter, I decided that no other theme could be more appropriate to the occasion.

It would be the merest naïvety to suppose that the idea of improvement – in its extreme or terminal form, of perfectibility – is a new one, or one that science has now authorized us to contemplate for the first time. The idea of improvement must be pretty well coeval with human speculative thought. In one form or another it embodies almost the whole spiritual history of mankind. No one man is qualified to give a complete account of the idea of improvement, even in the secular aspects to which I shall confine myself; but amidst all the profusion and confusion of thought on the subject it is, I think, possible to discern three main kinds of conception or vision about what the future of man in the world might be. I shall call them Olympian, Arcadian and Utopian.

In the Olympian conception, men can become like gods; can achieve complete virtue, understanding and peace of mind, but through spiritual insight, not by mastery of the physical world. No particular environment is envisaged as a setting for this apotheosis, and the environment itself need not have been perfected; the Olympian formula is for all seasons. In the Olympian vision, the direction of the human gaze is upwards or perhaps inwards; but in Arcadian thought, closely bound up with the ancient legend of a Golden Age, it is directed backwards. In Arcadia men remain human but in a state of natural innocence. They retreat into a tranquil pastoral world where peace of mind is not threatened, intellectual aspiration is not called for, and virtue is not at risk. An Arcadian society is anarchic; everything that is implied by authority is replaced by everything that is implied by fraternity. It is a world without strife, without ambition, and without material accomplishment.

From the seventeenth century onwards a new vision began to be taken seriously, the Utopian. Man can create anew and therefore improve the world he lives in through his own exertions; he begins as a tenant or lodger in the world, but ends up as its landlord; and as his environment improves, so, it is alleged, will he. Virtue can be learned and will eventually become second nature, understanding can be aspired to, but complete peace of mind can never be achieved because there will always be something more to do. Men look forwards, never backwards, and seldom upwards.

All three visions have both noble and comic elements, and each has developed its own satirical literature, which is often better known than the work it satirizes. In spite of these and other discouragements, scientists are characteristically Utopian in their outlook, because it is the only scheme of belief that makes sense of what scientists do. There is nevertheless one long recognized weakness in Utopian speculation: the inadequacy of man, the extreme unlikelihood that man can live up to his own ambitions. It is for this reason that the idea of a genetic improvement of man has a special fascination for Utopian thinkers. One of the three princes of Campanella's *The City of the Sun*, a Utopia of the early seventeenth century, is named Love, and Love's business is to supervise a system of eugenic mating. 'He sees that men and women are so joined together that they bring forth the best offspring', Campanella's narrator says: 'Indeed they laugh at us who exhibit a studious care for our breed of horses and dogs, but neglect the breeding of human beings.'

The idea of genetic improvement as a realizable policy may be said to have begun with the writings of Francis Galton, the great nineteenth-century humanist who founded the science of eugenics, and coined the word itself. 'Eugenics', said Galton, 'is the science which deals with all the influences that improve the inborn qualities of a race; also with those that develop them to the utmost advantage.' 'Man is gifted with pity and other kindly feelings; he has also the power of preventing many kinds of suffering. I conceive it to fall well within his province to replace natural selection by other processes that are more merciful and not less effective. That is precisely the aim of eugenics.' We must understand that when Galton defined eugenics in these terms, he was combating a radical form of social Darwinism, especially championed by Ernst Haeckel, according to which the doctrine of the survival of the fittest applied to the social development of human beings no less than to the evolution of animal communities. A quotation from Haeckel himself will make my point: 'The theory of selection teaches us that in human life, exactly as in animal and plant life, at each place and time only the small privileged minority can continue to exist and flourish; the great mass must starve and more or less prematurely perish in misery. . . . We may deeply mourn this tragic fact, but we cannot deny or alter it.'

Social Darwinism in the form expounded by Haeckel provided a theoretical justification for the great biological crimes of Fascism, so it is hardly surprising that eugenics fell into complete discredit. Politically speaking its object was (I quote Condorcet) 'to render Nature herself an accomplice in the guilt of political inequality'. In spite of this, kindly and humane people, genuine descendants of Galton himself, have continued to believe that genetic policies can be used to improve the performance and capabilities of human beings, and that our power to use genetics for this purpose offers an antidote to the slow deterioration of man thought to be produced by the amelioration or softening of the environment, and is the foundation of all rational hope for the improvement of human society.

The case for 'positive eugenics', that is for constructive rather than merely remedial eugenics, is based on the model of stockbreeding. If horses, dogs and cattle can be improved by selective breeding, it is argued, why cannot human beings? (This is Campanella's question.)

One can give two kinds of answer to this question – a moral and political answer, and a scientific answer. The moral-political answer is that no such regimen of genetic improvement could be practised within the framework of a society that respects the rights of individuals. This answer should be sufficient, but I am going to give the scientific answer also, for the following very important reason. Many people believe that it is scientifically feasible to create a population of supermen, and that only man's better self stands in the way of putting such a policy into practice. This belief, I hope to show, is quite erroneous, but it is important to refute it, because it is the kind of belief that underlies the modern conception of science as an essentially dehumanizing activity which may perpetrate some terrible mischief if it is not kept firmly under a control which it is at all times striving to elude.

In point of fact science has very little to do with it. The empirical arts of the stockbreeder are as old as civilization. Given a tyrant or a dynasty of tyrants, a scheme of selective inbreeding could have been enforced upon human beings at any time within the past five thousand years. It is not something that science has now for the first time put it within our power to do. At any time in the past few thousand years it would have been possible to embark on a scheme to make human beings as different one from another in appearance and capabilities as greyhounds and pekinese.

Let me now try to explain why a regimen of selective inbreeding is not scientifically acceptable, and why the stockbreeding analogy can no longer be sustained. In the old days (not so very long ago) the end-products of the stockbreeder's art – whether super-dogs or super-cattle or even super-mice – were expected to fulfil not one but *two* functions: two functions which, until recently, no one clearly distinguished and no one clearly realized could be separately and independently fulfilled. The first function was to be the end-product itself, to be the usable, eatable, or marketable goal of the breeding procedure. The second function was to be the parents of the next generation of super animals. In order to fulfil this second or reproductive function, the eugenic end-product had to meet a certain genetic specification, viz. that it should be homozygous or true-breeding in respect of all the characters for which selection had been exercised. If the animals were predominantly heterozygous or genetically mixed in their make-up, then the stock would not breed true and the efforts of the stockbreeder would be dissipated in a single generation.

Until recent years the dual ambition of the eugenic stockbreeder seemed to be upheld by genetic theory. The prevailing belief was that the animals belonging to a particular species or interbreeding community were predominantly homozygous in genetic composition. Natural selection was thought to be working towards the establishment or fixation of a particular genotype or genetic make-up, that which conferred the highest degree of adaptedness to the prevailing circumstances. If the circumstances changed, so also would the genetic make-up, because new genes – mutant genes – were continually proffered for selection, and these provided resources of variation rich enough to make it possible for natural selection to work out a new and improved genetic formula for survival. It is true that polymorphism was recognized as a departure from this tidy scheme (polymorphism refers to the stable co-existence of genetically differentiated types in the population for reasons other than the pressure of recurrent mutations); but polymorphism was thought of as a special phenomenon for the existence of which special explanations had to be devised. By and large, nevertheless, the stockbreeder and the theoretical geneticist supported each other's conceptions: genetic theory made sense of the stockbreeder's ambitions, and the stockbreeder could think of himself as a man who put genetic theory to practical use.

Today and for some years past this entire scheme of thinking has been called into question. It turns out that natural populations of outbreeding organisms, including human beings, are persistently and obstinately diverse in genetic make-up. Polymorphism is not an exceptional phenomenon but the rule. One drop of human blood can tell an astonishing story of human diversity. The haemoglobins and non-haemoglobin proteins, the red cell anti-gens, the serum enzymes and serum proteins generally, the leuco-cyte surface structures, all exist in a huge profusion of variant forms; and the same is almost certainly true of all the other macro-molecular constituents of the body. People have therefore come to abandon the view that natural selection works towards the fixation of a particular genotype, of some one preferred genetic formula for adaptedness. It is *populations* that evolve, not pedi-grees; and the end-product of evolution, in so far as it can be said to have one, is itself a population, not a representative genetic type to which every individual will represent a more or less faithful approximation. The individual members of the population differ from one another, but the population itself has a stable genetic structure, i.e. a stable pattern of genetic inequality. Individual members of the population do not breed true, for being hetero-zygous their offspring will necessarily be unlike themselves. But, as G. H. Hardy pointed out more than sixty years ago, the Mendelian process is such that the population as a whole breeds true – i.e. reproduces a population of the same genetic structure as itself – even if its individual members do not. Under a regimen of random mating, the frequency with which the various genotypes appear in the population remains constant from generation to generation, except in so far as natural selection may change their proportions, and so cause a new pattern of genetic inequality to take shape. 'Natural selection' in this context refers simply to a state of affairs in which the different genetic categories do not make an equal contribution to the ancestry of future generations, i.e. a contribu-tion proportional to their existing numbers. Some take a larger share and others necessarily a smaller share, and so the genetic make-up of the population changes.

This newer conception undermines the ambition of the old-fashioned stockbreeder, and makes nonsense of the eugenic ambitions that seemed to be supported by their practice. The stockbreeder is now seen to be undertaking an unnatural pro-cedure, no longer authorized by our conception of the way that genetic changes happen in real life.

It is now known that this dilemma can be resolved, at least in principle. The stockbreeder must now no longer expect his animals to fulfil both the functions which, he supposed, went necessarily together: they cannot both represent the eugenic end-product *and* be the parents of succeeding generations.

To a layman the idea of dissociating the existential and reproductive functions of the eugenic end-product seems impossible to achieve except by some sort of conjuring trick – certainly impossible to reconcile with having eugenic end-products that are uniform in the characteristics they have been bred to possess. The trick is done, of course, by adopting a nicely calculated regimen of cross-breeding.

The principle is simple enough. Let me illustrate it from the world of laboratory animals. If I maintain two inbred strains of mice in my laboratory – strains sufficiently inbred to be homozygous and therefore true-breeding at most genetic loci – the animals representing the first generation of a cross between them will also form a uniform population. They will also probably be better performers than their parents in every way: in fertility, growth rate, intelligence, longevity, and resistance to disease. The characteristics of these first generation hybrids are specified by the genetic properties of the parental strains, and if the parents are in fact judiciously chosen, the hybrids may represent the eugenic end-product which we seek. They are, however, hybrid or heterozygous for every gene in respect of which the two parental strains differ. Therefore they cannot be bred from, because their own offspring will be not only diverse but maximally diverse, in the sense that they will exploit to the full the genetic possibilities defined by the make-up of the two parental strains. If therefore I wanted to raise a uniform population of super-mice at will, I should try to breed two homozygous strains whose F1 hybrid progeny answered to my specifications, but they themselves, the progeny, would be relieved of a reproductive function, which would continue to be exercised by the two parental strains. Thus the dissociation between being the eugenic end-product and being the parents of the next generation would be complete.

Modern stockbreeding practice makes use of very much more sophisticated schemes of cross-breeding than this, but the principle is the same. The somatic and reproductive functions are separated: the eugenic end-product is reproducible at will, but does not reproduce.

This concludes my statement of the scientific reasons why the goal of positive eugenics, as Galton and his followers envisaged it, cannot be achieved. Human diversity is one of the facts of life, and the human genetic system does not lend itself to improvement by selective inbreeding. We could not adapt modern stockbreeding principles to a human society without abandoning a large part of what we understand by being human. Although negative or purely remedial eugenics has a useful and important function to fulfil in human society, I think that, in the main, for many centuries to come, we shall have to put up with human beings as they are at present constituted.

Animal Experimentation in a Medical Research Institute[1]

The Stephen Paget Lecture has as its particular theme a defence of the use of experimental animals to enlarge medical knowledge. You may well wonder why in the year 1966 such a defence should be thought necessary, and, conversely, why the general public should demand repeated assurances that medical research is being humanely and properly conducted. I myself believe it is entirely right that the public should ask for these assurances. When I say 'properly conducted' I do not mean properly conducted *only* in respect of experiments on animals (although that happens to be the particular theme of this lecture), but properly conducted in respect of all research activities that could reasonably cause misgivings. For example, the possible dangers of clinical experimentation; the endeavour to keep people alive beyond what is thought to be their natural span by the use of medical contrivances of one kind or another – or alternatively, the morality of *not* keeping them alive when it is in principle possible to do so. Then there are the possible dangers of our great and ever-growing dependence on medical supplies and medical services, a dependence so great as to tempt people to say that one day the whole world will turn into a kind of hospital in which even the best of us will be no more than ambulatory patients; and the dangers, real or imagined, of the genetic deterioration brought about by the propagation of the genetically unfit.

Some of these dangers are illusory, and can be shown to be so; but the fears and misgivings they give rise to are not illusory, and they must be allayed – by public discussion, by making the truth of these matters widely known, and by such methods as the delivery of Stephen Paget Memorial Lectures.

[1] The National Institute for Medical Research of which I was director 1962–1971.

In saying that medical research workers should be required to give a fair account of themselves to the general public, I am talking as if the general public were a sort of all-wise body into whose care the well-being of animals could perfectly safely be entrusted. Alas – this is very far from being the case. The general public is by no means qualified to judge whether or not our human wardenship of animals is being satisfactorily discharged.

Some years ago I had the privilege of serving on a Home Office Committee 'to inquire into practices or activities which may involve cruelty to British wild mammals, whether at large or in captivity'. We took great pains to hear evidence from all interested parties, but the amount of evidence that bore on the welfare of unattractive animals, or on pests like rats, was negligible.

It is difficult not to despise the sentimental ignorance about animals that is so widely thought of as a traditional part of the British character – the kind of ignorant sentimentality that finds expression in the fatuous cry that a caged bird should be 'given its freedom'. Somebody should make the general public familiar with modern research on the dynamics of natural populations of animals: for example, the work in which Professor Lack has shown that the annual adult mortality of the European robin is as high as 60 per cent, of the starling 50 per cent and of the sparrow no less than 45 per cent. The concern of the British public for the welfare of animals is, as a matter of fact, a rather new thing: it does not lie deep in our traditions. I think I am right in saying that the common law takes no cognizance of the rights of animals, and I do not know if it even concedes that animals can have rights. At all events, the first legislation protecting animals dates from the 1820s (the RSPCA was founded in 1824). The reason given for introducing new legislation to prohibit cock fighting was that it tended to corrupt the general public – not that it inflicted cruelty on the wretched animals themselves. Although I disapprove of pop sociology, a good case can be made for arguing that ignorant sentimentality about animals and how they live in nature grew up in proportion as people ceased to know anything about animals at first hand. The literature personalizing animals has grown up in the past hundred years. *Alice in Wonderland* was published in 1865 and *Black Beauty* in 1877, and soon the nursery came to be populated with animal familiars – Brer Rabbit and Peter Rabbit, Piglet and Donald Duck have been the conditioning stimuli of our childhood; but we must allow ourselves to grow up if we are to get any sensible conception of the nature and life of animals as they really are.

The opposite of ignorant sentimentality is humane understanding. Just how far the public has yet to go to achieve humane understanding is made very clear by the world of pet dogs and the Dog Shows.

During the past ten years or so, the British Veterinary Association – in particular the Small Animals Veterinary Association – and the Animal Health Trust have been fighting an uphill but in the main successful battle to educate dog breeders and their clients, show judges and the 650-odd Breed Societies, into some understanding of, and some determination to cope with, the problems raised by the occurrence in many breeds of dogs of distressing or painful congenital deformities. I remember being stirred by a Presidential Address on this very theme at the Annual Meeting of the British Veterinary Association in 1954.

The Kennel Club has co-operated with the veterinary profession in the exposure and analysis of these abnormalities, and the fruits of their co-operation can be read in a series of important papers in the *Journal of Small Animal Practice*. They make sorry reading. The gist of them is that most breeds of dogs carry a cruel load of abnormalities which are the primary or secondary consequences of hereditary defects. Among them are dislocation of the knee cap or hip, gross skeletal deformities, undescended testes, deafness, retinal atrophy, chronic dermatitis, ingrowing eyelashes and chronic respiratory distress; they extend even to hyperexcitability, mental deficiency, or downright idiocy.

Now these deformities are of two kinds. Some are quite unwanted, and are indeed accidental. They have been unluckily fixed by inbreeding, and they remain in the stocks because breeders are more anxious to sell and unwilling to cull. These abnormalities are not approved of, but they are condoned. Other congenital abnormalities are deliberately bred for; they are show points; they are among the defining characters of the breed. I do not understand how anyone of educated sensibility can admire the bow legs and poor crumpled face of the bulldog, the spinal deformity that gives him his gay twirly tail, the palatal abnormalities that make it so difficult for him to breathe. No one with a real understanding of animals could applaud a show stance made possible by a congenital dislocation of the hip. We should all applaud the British Veterinary Association and the Animal Health Trust for the stand they have taken; and let me add that their criticism of breeders and of show judges was very much more warmly expressed than my own.

The welfare of animals must depend on an *understanding* of animals, and one does not come by this understanding intuitively: it must be learned. I once knew a little girl,[1] who having been told that frogs were rather engaging creatures, befriended a frog. Her first thought was that it needed warming up, because it felt so cold. Her first lesson in the humane understanding of animals was that frogs do not like being warmed up and do better at the temperature of their environment.

Fortunately, some humane and learned organizations do exist to promote the welfare of animals and to educate the public to understand animals as they really are, so that they need no longer rely on some supposedly intuitive understanding of what animals think or feel. One of the most important of these organizations is UFAW, the Universities Federation for Animal Welfare. I had the pleasure of being the Chairman of its Scientific Advisory Committee for some few years. Among these organizations I include that department of the Home Office which authorizes and supervises all experimentation on animals in Institutes such as my own. Not everyone realizes what a high proportion of medical research workers in this country do warmly approve of the restriction of animal experimentation to people who are qualified to carry it out. This is not to say that the existing Home Office

[1] A dedicatee of this book.

regulations are flawless or that certain administrative changes in their working are not now widely thought to be desirable.

After this long preamble – I intended it to be so – let me now say something about the care of and the use of animals in the National Institute for Medical Research, the largest research institute of its kind in the Commonwealth.

The NIMR is a sort of microcosm of basic medical research.

So far as is possible or practicable, the animals used in the Institute are bred within its precincts. We like to think of ourselves as the pioneers in this country of the careful and the scientifically informed husbandry of laboratory animals. The animals are in charge of a scientific division of the Institute headed by a senior veterinary scientist. The Superintendent of the Division is responsible not only for its day-to-day running, but also for the training of animal technicians – educating them for a career which, thanks to Mr D. J. Short's efforts as much as to anybody's, now offers the prospect of a rewarding and interesting life in what has come to be thought of as a profession ancillary to medical science. The establishment of animal technicians as a recognized profession and the regulation of examination standards by an Institute were projects in which the National Institute for Medical Research is proud to have played a leading part.

This was a revolution, for in the old days the care of animals was too often entrusted to kindly and well meaning, but often not very bright, old men. There have been two other such revolutions in laboratory animal husbandry. The second was the provision and use of animals of known genetic composition and history – notably of inbred animals and of first generation hybrids between inbred strains. This innovation met with a good deal of opposition from the medical profession. It was contended that pure bred animals were in some way artificial and unnatural, and that the results secured by using them would be misleading or unrepresentative. This criticism is, of course, based on a misunderstanding of the nature and purposes of medical research; one might with equal justice reproach the chemist for basing his researches on the use of pure compounds.

F

The third revolution, which is still in progress, is the control of infectious disease. The animals in Research Institutes are of necessity kept at a population density which makes them an easy prey to epidemics. To control these epidemics – or rather, to prevent their occurrence in the first instance – is essentially a problem in medical or sanitary engineering. The same is true of the control of epidemics in human populations. Indeed, the actuarial status of the experimental animals bred in almost all laboratories today is still too much like that of a human population in the sixteenth century: a grievously high proportion of the animals still die from intercurrent infection, and without the consolation of believing that they go to heaven. The principle of protecting animal colonies from the attacks of pathogenic organisms had also to be fought for, but the battle has been won, and within the next few years all major biological Research Institutions will re-found their animal colonies on what is called a 'specific pathogen free' (SPF) basis.

In the National Institute for Medical Research, as in the country generally, the largest single users of experimental animals are those responsible for the standardization and safety control of drugs and vaccines. The National Institute is in fact an agent of the World Health Organization for defining international standards of those drugs and biological products which can only be assayed and tested by biological methods; and we are also agents of the Ministry of Health for checking the safety of vaccines and other agents of an immunological nature used in general medical practice. The two scientific divisions of the Institute responsible for this work make use of about 25,000 experimental animals a year; the control of polio vaccines makes use of some 2,000 monkeys a year. It is a large number in any absolute sense, but very small in proportion to the number of children at risk. Of course, no one is satisfied with the use of experimental animals for these purposes. The most determined efforts are constantly being made to find substitutes for the use of animals in standardization and control. To describe the directions that research is taking, I cannot improve on Dr W. M. S. Russell's '3 Rs' of humane laboratory practice; Reduction, Refinement, and Replacement. The number of animals used may be reduced only by increasing the amount of information to be secured from the study of any one. One method of doing so is to use genetically standardized animals – not necessarily inbreds. 'Refinement' is essentially a matter of increasing the precision of individual assays. Our ultimate goal, however, is the replacement of animals altogether. For drugs, we look forward ultimately to chemical assays, or at least to the adoption of *in vitro* methods, such as the immunological assays now being developed for the standardization of protein hormones. For the safety control of virus vaccines, e.g. polio vaccine, everyone hopes that the cytopathic changes produced in cells in tissue cultures will prove to be sufficiently discriminating and reliable. Slowly but progressively all these ambitions are being achieved.

These activities of the Institute are services, though they are services underpinned by research. Turning now to the researches of the Institute generally, I obviously cannot describe the dozens of projects undertaken in this past year by 180 scientists supported by perhaps twice that number of qualified technicians, but I shall choose some special examples to give you some idea of their variety and range of purposes.

Some people believe that the greatest task of modern medicine is to extend to the world generally the standards of medicine and hygiene that today obtain only in the advanced industrial countries. Let me therefore first mention the Institute's researches into malaria and leprosy in collaboration with research stations in West Africa and in Malaya respectively. Malaria is still, in a numerical sense, the world's gravest disease: some two and a half million people die every year of malaria and perhaps two hundred and fifty million are afflicted by it at any one time so it is not unreasonable that less than one-thousandth of that number of animals should be used in experimental malarial research. Rats and monkeys are each susceptible to their own kinds of malarial infection, and the study of rats and monkeys has already taken us a long way towards elucidating the mechanism of the cyclical recurrence of malarial illness. Our understanding of leprosy is still backward, because the organism that causes it – it belongs to the same family as the tubercle bacillus – cannot yet be grown in cell-free cultures, and grows impossibly slowly when caused to infect cells in tissue culture; but rodents can be infected with their own leprosy organism, and now at last it has become possible to grow the human organism in mice. Now, for the first time, critical experimental tests on the chemotherapy of leprosy can be undertaken.

My own special research interest is in the field of transplantation, and our ambition is to overcome the immunological barriers that normally prohibit the transplantation of tissues and organs between two different human beings or, for that matter, two different mice. The transplantation of kidneys in medical practice has already enjoyed greater success than any of us dared to believe possible even as recently as five years ago. All the methods used in clinical practice to prolong the life of homografts have been founded upon experiments carried out in mice, rabbits, rats and dogs, and thanks to them, surgery will one day enter into that new dimension of accomplishment which the transplantation of organs seems to promise.

The transplantation problem is a problem in immunology. Our Institute has been described as the greatest centre of immunological research in the world, and I shall not challenge this description. A high proportion of immunological research is now directed towards inhibiting and controlling the immunological response. When that control has been achieved, as it certainly will be, its rewards will be diffused far more widely than over the domain of transplantation itself. It will become possible to relieve that huge diffuse burden of human suffering imposed upon us by the allergies, hypersensitivities, auto-immune diseases, and many other miscarriages of the immunological process.

Research at the Institute is by no means confined to the use of lower animals; perhaps no Institute makes greater use of human volunteers for those researches in which only human beings will do. The complex of viruses responsible for the common cold – viruses first defined and cultivated in the National Institute – cause their characteristic symptoms only in man. Human volunteers are therefore used at the Common Cold Research Unit, our famous outpost near Salisbury. Human volunteers are also, of necessity, used to study the adaptation of human beings to climatic stresses. The Hampstead campus of the Institute is equipped with the complex instrumentation that makes it possible to create climatic conditions even more disagreeable than those which prevail out-of-doors. One of our most distinguished human physiologists in Hampstead is studying the athletic performance of human beings at an altitude of 7,415 ft. above sea level – the height of Mexico City, where the Olympic Games were held in 1968. It would not be very informative to simulate the Olympic Games with mice.

You may think that in choosing malaria, leprosy, and transplantation as examples of the Institute's research, I am cheating – at least in the sense that I am directing your attention to research of obvious practical utility, where there is no doubting the ultimate benefit to mankind. But what about the moral credentials of so called 'pure' research – for example on the mechanisms of protein synthesis, one of our major preoccupations?

In terms of their ultimate relevance to mankind, the difference between research into protein synthesis and on malaria is a difference of immediacy, and in the degree of diffusion of their effects. The work on protein synthesis stands further from practical application than work on malaria, but its results illuminate almost the whole of biology and medicine; they illuminate normal and abnormal growth, regeneration, reproduction, the synthesis of hormones, the production of antibodies, the multiplication of viruses and bacteria – surely a big enough dividend for any scholarly investment. You may disapprove of *all* experiments using animals, but it is scientifically and medically ruinous to approve only those with obvious practical uses and to reprobate all others.

To conclude: the use of experimental animals in medical research requires justification, and I think that the general public is right to demand repeated assurances that such a justification exists. The justification lies in the advancement of human welfare, but I myself interpret 'welfare' more widely than in terms of material benefits or the conquest of disease. Human beings are so constituted that they seem temperamentally obliged to explore the world around them, to enlarge their grasp and understanding of nature. It is to this restless and insistent exploratory process that human beings owe their present place in the world. It is too late now to adopt an intellectually pastoral existence – to adopt a molluscan solution of the problems of living. To invert an epigram of Thomas Browne's, it is too late to cease to be ambitious. The use of experimental animals in laboratories to enlarge our understanding of nature is part of a far wider exploratory process, and one cannot assay its value in isolation – as if it were an activity which, if prohibited, would deprive us only of the material benefits that grow directly out of its own use. Any such prohibition of learning or confinement of the understanding would have widespread and damaging consequences; but this does not imply that we are for evermore, and in increasing numbers, to enlist animals in the scientific service of man. I think that the use of experimental animals on the present scale is a temporary episode in biological and medical history, and that its peak will be reached in ten years time, or perhaps even sooner. In the meantime we must grapple with the paradox that nothing but research on animals will provide us with the knowledge that will make it possible for us, one day, to dispense with the use of them altogether.

Science and the Sanctity of Life

I do not intend to deny that the advances of science may sometimes have consequences that endanger, if not life itself, then the
quality of life or our self-respect as human beings (for it is in this
wider sense that I think 'sanctity' should be construed). Nor shall
I waste time by defending science as a whole or scientists generally
against a charge of inner or essential malevolence. The Wicked
Scientist is not to be taken seriously: Dr Strangelove, Dr Moreau,
Dr Moriarty, Dr Mabuse, Dr Frankenstein (an honorary degree,
this), and the rest of them are puppets of Gothic fiction. Scientists,
on the whole, are amiable and well-meaning creatures. There
must be very few wicked scientists. There are, however, plenty of
wicked philosophers, wicked priests, and wicked politicians.

One of the gravest charges ever made against science is that
biology has now put it into our power to corrupt both the body
and the mind of man. By scientific means (the charge runs) we can
now breed different kinds and different races – different 'makes',
almost – of human beings, degrading some, making aristocrats of
others, adapting others still to special purposes: treating them in
fact like dogs, for this is how we *have* treated dogs. Or again:
science now makes it possible to dominate and control the thought
of human beings – to improve them, perhaps, if that should be our
purpose, but more often to enslave or to corrupt with evil teaching.

But these things have always been possible. At any time in the
past five thousand years it would have been within our power to
embark on a programme of selecting and culling human beings
and raising breeds as different from one another as toy Poodles and
Pekinese are from St Bernards and Great Danes. In a genetic sense
the empirical arts of the breeder are as easily applicable to human
beings as to horses – more easily applicable, in fact, for human
beings are highly *evolvable* animals, a property they owe partly to

an open and uncomplicated breeding system, which allows them a glorious range of inborn diversity and therefore a tremendous evolutionary potential; and partly to their lack of physical specializations (in the sense in which ant-eaters and woodpeckers and indeed dogs are specialized), a property which gives human beings a sort of amateur status among animals. And it has always been possible to pervert or corrupt human beings by coercion, propaganda, or evil indoctrination. Science has not yet improved these methods, nor have scientists used them. They have, however, been used to great effect by politicians, philosophers and priests.

The mischief that science may do grows just as often out of trying to do good – as, for example, improving the yield of soil is intended to do good – as out of actions intended to be destructive. The reason is simple enough: however hard we try, we do not and sometimes cannot foresee all the distant consequences of scientific innovation. No one clearly foresaw that the widespread use of antibiotics might bring about an evolution of organisms resistant to their action. No one could have predicted that X-irradiation was a possible cause of cancer. No one could have foreseen the speed and scale with which advances in medicine and public health would create a problem of overpopulation that threatens to undo much of what medical science has worked for. (Thirty years ago the talk was all of how the people of the Western world were reproducing themselves too slowly to make good the wastage of mortality; we heard tell of a 'Twilight of Parenthood', and wondered rather fearfully where it all would end.) But somehow or other we shall get round all these problems, for every one of them is soluble, even the population problem, and even though its solution is obstructed above all else by the bigotry of some of our fellow men.

I choose from medicine and medical biology one or two concrete examples of how advances in science threaten or seem to threaten the sanctity of human life. Many of these threats, of course, are in no sense distinctively medical, though they are often loosely classified as such. They are merely medical contexts for far more pervasive dangers. One of them is our increasing state of dependence on medical services and the medical industries. What would become of the diabetic if the supplies of insulin dried up, or of the victims of Addison's disease deprived of synthetic steroids? Questions of this kind might be asked of every service society provides. In a complex society we all sustain and depend upon each other – for transport, communications, food, goods, shelter, protection, and a hundred other things. The medical industries will not break down all by themselves, and if they do break down it will be only one episode of a far greater disaster.

The same goes for the economic burden imposed by illness in any community that takes some collective responsibility for the health of its citizens. All shared burdens have a cost which is to a greater or lesser degree shared between us: education, pensions, social welfare, legal aid, and every other social service, including government.

We are getting nearer what is distinctively medical when we ask ourselves about the economics, logistics, and morality of keeping people alive by medical intervention and medical devices. At present it is the cost and complexity of the operation, and the shortage of machines and organs, that denies a kidney graft or an artificial kidney to anyone mortally in need of it. The limiting factors are thus still economic and logistic. But what about the morality of keeping people alive by these heroic medical contrivances? I do not think it is possible to give any answer that is universally valid or that, if it were valid, would remain so for more than a very few years. Medical contrivances extend all the way from pills and plasters and bottles of tonic to complex mechanical prostheses, which will one day include mechanical hearts. At what point shall we say we are wantonly interfering with Nature and prolonging life beyond what is proper and humane?

In practice the answer we give is founded not upon abstract moralizing but upon a certain natural sense of the fitness of things, a feeling that is shared by most kind and reasonable people even if we cannot define it in philosophically defensible or legally accountable terms. It is only at international conferences that we tend to adopt the convention that people behave like idiots unless acting upon clear and well-turned instructions to behave sensibly. There is in fact no general formula or smooth form of words we can appeal to when in perplexity.

Moreover, our sense of what is fit and proper is not something fixed, as if it were inborn and instinctual. It changes as our experience grows, as our understanding deepens, and as we enlarge our grasp of possibilities – just as living religions and laws change, and social structures and family relationships.

I feel that our sense of what is right and just is already beginning to be offended by the idea of taking great exertions to keep alive grossly deformed or monstrous newborn children, particularly if their deformities of body or mind arise from major defects of the genetic apparatus. There are in fact scientific reasons for changing an opinion that might have seemed just and reasonable a hundred years ago.

Everybody takes it for granted, because it is so obviously true, that a married couple will have children of very different kinds and constitutions on different occasions. But the traditional opinion, which most of us are still unconsciously guided by, is that the child conceived on any one occasion is the unique and necessary product of that occasion: *that* child would have been conceived, we tend to think, or no child at all. This interpretation is quite false, but human dignity and security clamour for it. A child sometimes wonderingly acknowledges that he would never have been born at all if his mother and father had not chanced to meet and fall in love and marry. He does not realize that, instead of conceiving him, his parents might have conceived any one of a hundred thousand other children, all unlike each other and unlike himself. Only over the past one hundred years has it come to be realized that the child conceived on any one occasion belongs to a vast cohort of Possible Children, any one of whom might have been conceived and born if a different spermatozoon had chanced to fertilize the mother's egg cell – and the egg cell itself is only one of very many. It is a matter of luck then, a sort of genetic lottery. And sometimes it is cruelly bad luck – some terrible genetic conjunction, perhaps, which once in ten or twenty thousand times will bring together a matching pair of damaging recessive genes. Such a misfortune, being the outcome of a random process, is, considered in isolation, completely and essentially pointless. It is not even strictly true to say that a particular inborn abnormality must have lain within the genetic potentiality of the parents, for the malignant gene may have arisen *de novo* by mutation. The whole process is unhallowed – is, in the older sense of that word, profane.[1]

I am saying that if we feel ourselves under a moral obligation to make every possible exertion to keep a monstrous embryo or new born child alive *because* it is in some sense the naturally intended – and therefore the unique and privileged – product of its parents' union at the moment of its conception, then we are making an elementary and cruel blunder: for it is *luck* that determines which one child is in fact conceived out of the cohort of Possible Children that might have been conceived by those two parents on that occasion. I am not using the word 'luck' of conception as such, nor of the processes of embryonic and foetal growth, nor indeed in any sense that derogates from the wonder and awe in which we hold processes of great complexity and natural beauty which we do not fully understand; I am simply using it in its proper sense and proper place.[1]

This train of thought leads me directly to eugenics – 'the science', to quote its founder, Francis Galton, 'which deals with all the influences that improve the inborn qualities of a race; also with

---

[1] An eminent theologian once said to me that I was making altogether too much fuss about this kind of mischance. It was, he said, all in the nature of things and already comprehended within our way of thinking; it was not different in principle from being accidentally struck on the head by a falling roof tile. But I think there *is* an important difference of principle. In the process by which a chromosome is allotted to one germ cell rather than another and in the union of germ cells luck is of the very essence. The random element is an integral, indeed a defining characteristic of Mendelian inheritance. All I am saying is that it is difficult to wear a pious expression when the fall of the dice produces a child that is structurally or biochemically crippled from birth or conception.

[2] There are, perhaps, weighty legal and social reasons why even tragically deformed children should be kept alive (for who is to decide? and where do we draw the line?), but these are outside my terms of reference.

those that develop them to the utmost advantage.' Because the upper and lower boundaries of an individual's capability and performance are set by his genetic make-up, it is clear that if eugenic policies were to be ill-founded or mistakenly applied they could offer a most terrible threat to the sanctity and dignity of human life. This threat I shall now examine.

Eugenics is traditionally subdivided into positive and negative eugenics. Positive eugenics has to do with attempts to improve human beings by genetic policies, particularly policies founded upon selective or directed breeding. Negative eugenics has the lesser ambition of attempting to eradicate as many as possible of our inborn imperfections. The distinction is useful and pragmatically valid for the following reasons.[1] Defects of the genetic constitution (such as those which manifest themselves as mongolism, haemophilia, galactosemia, phenylketonuria, and a hundred other hereditary abnormalities) have a much simpler genetic basis than desirable characteristics like beauty, high physical performance, intelligence, or fertility. This is almost self-evident. All geneticists believe that 'fitness' in its most general sense depends on a nicely balanced co-ordination and interaction of genetic factors, itself the product of laborious and long drawn out evolutionary adjustment. It is inconceivable, indeed self-contradictory, that an animal should evolve into the possession of some complex pattern of interaction between genes that made it inefficient, undesirable, or unfit – i.e. *less* well adapted to the prevailing circumstances. Likewise, a motor-car will run badly for any one of a multitude of particular and special reasons, but runs well because of the harmonious mechanical interactions made possible by a sound and economically viable design.

Negative eugenics is a more manageable and understandable enterprise than positive eugenics. Nevertheless, many well-meaning people believe that, with the knowledge and skills already available to us, and within the framework of a society that upholds the rights of individuals, it is possible in principle to raise a superior kind of human being by a controlled or 'recommended'

[1] See my book *The Future of Man* (Methuen, 1960).

scheme of mating and by regulating the number of children each couple should be allowed or encouraged to have. If stockbreeders can do it, the argument runs, why should not we? – for who can deny that domesticated animals have been improved by deliberate human intervention?

I think this argument is unsound for a lesser and for a more important reason.

1. Domesticated animals have not been 'improved' in the sense in which we should use that word of human beings. They have not enjoyed an all-round improvement, for some special characteristics or faculties have been so far as possible 'fixed' without special regard to and sometimes at the expense of others. Tameness and docility are most easily achieved at the expense of intelligence, but that does not matter if what we are interested in is, say, the quality and yield of wool.

2. The ambition of the stockbreeder in the past, though he did not realize it, was twofold: not merely to achieve a predictably uniform product by artificial selection, but also to establish an internal genetic uniformity (homozygosity) in respect of the characters under selection to make sure that the stock would 'breed true' – for it would be a disaster if characters selected over many generations were to be irrecoverably lost or mixed up in a hybrid progeny. The older stockbreeder believed that uniformity and breeding true were characteristics that necessarily went together, whereas we now know that they can be separately achieved. And he expected his product to fulfil two quite distinct functions which we now know to be separable, and often better separated: on the one hand, to be in themselves the favoured stock and the top performers – the super-sheep or super-mice – and, on the other hand, to be the parents of the next generation of that stock. It is rather as if Rolls-Royces, in addition to being an end-product of manufacture, had to be so designed as to give rise to Rolls-Royce progeny.[1]

It is just as well these older views are mistaken, for with naturally

[1] These arguments are set out more fully in 'The Genetic Improvement of Man', pp. 69–76.

outbreeding populations such as our own, genetic uniformity, arrived at and maintained by selective inbreeding, is a highy artificial state of affairs with many inherent and ineradicable disadvantages.

Stockbreeders, under genetic guidance, are now therefore inclining more and more towards a policy of deliberate and nicely calculated cross-breeding. In the simplest case, two partially inbred and internally uniform stocks are raised and perpetuated to provide two uniform lineages of parents, but the eugenic goal, the marketable end-product or high performer, is the progeny of a cross between members of the two parental stocks. Being of hybrid make-up, the progeny do not breed true, and are not in fact bred from; they can be likened to a manufactured end-product; but they can be uniformly reproduced at will by crossing the two parental stocks. Many more sophisticated regimens of cross-breeding have been adopted or attempted, but the innovation of principle is the same. (1) The end-products are all like each other and are faithfully reproducible, but are not bred from because they do not breed true: the organisms that represent the eugenic goal have been relieved of the responsibility of reproducing themselves. And (2) the end-products, though uniform in the sense of being like each other, are to a large extent hybrid – heterozygous as opposed to homozygous – in genetic composition.

The practices of stockbreeders can therefore no longer be used to support the argument that a policy of positive eugenics is applicable in principle to human beings in a society respecting the rights of individuals. The genetical manufacture of supermen by a policy of cross-breeding between two or more parental stocks is unacceptable today, and the idea that it might one day become acceptable is unacceptable also.

A deep fallacy does in fact eat into the theoretical foundations of positive eugenics and that older conception of stockbreeding out of which it grew.[1] The fallacy was to suppose that the *product* of

[1] See my article, 'A Biological Retrospect', *Nature* (London, 25 September 1965), vol. 207, p. 1327.

evolution, i.e. the outcome of an episode of evolutionary change, was a new and improved genetic formula (genotype) which conferred a higher degree of adaptedness on the individuals that possessed it. This improved formula, representing a new and more successful solution of the problems of remaining alive in a hostile environment, was thought to be shared by nearly all members of the newly evolved population, and to be stable except in so far as further evolution might cause it to change again. Moreover, the population would have to be predominantly homozygous in respect of the genetic factors entering into the new formula, for otherwise the individuals possessing it would not breed true to type, and everything natural selection had won would be squandered in succeeding generations.

Most geneticists think this view mistaken. It is *populations* that evolve, not the lineages and pedigrees of old-fashioned evolutionary 'family trees', and the end-product of an evolutionary episode is not a new genetic formula enjoyed by a group of similar individuals, but a new spectrum of genotypes, a new pattern of genetic inequality, definable only in terms of the population as a whole. Naturally outbreeding populations are not genetically uniform, even to a first approximation. They are persistently and obstinately diverse in respect of nearly all constitutive characters which have been studied deeply enough to say for certain whether they are uniform or not. It is the *population* that breeds true, not its individual members. The progeny of a given population are themselves a population with the same pattern of genetic make-up as their parents – except in so far as evolutionary or selective forces may have altered it. Nor should we think of uniformity as a desirable state of affairs which *we* can achieve even if nature, unaided, cannot. It is inherently undesirable, for a great many reasons.

The goal of positive eugenics, in its older form, cannot be achieved, and I feel that eugenic policy must be confined (paraphrasing Karl Popper) to *piecemeal genetic engineering*. That is just what negative eugenics amounts to; and now, rather than to deal in generalities,

I should like to consider a concrete eugenic problem and discuss the morality of one of its possible solutions.

Some 'inborn' defects – some defects that are the direct consequence of an individual's genetic make-up as it was fixed at the moment of conception – are said to be of *recessive* determination. By a recessive defect is meant one that is caused by, to put it crudely, a 'bad' gene that must be present in *both* the gametes that unite to form a fertilized egg, i.e. in both spermatozoon and egg cell, not just in one or the other. If the bad gene *is* present in only one of the gametes, the individual that grows out of its fusion with the other is said to be a *carrier* (technically, a heterozygote).

Recessive defects are individually rather rare – their frequency is of the order of one in ten thousand – but collectively they are most important. Among them are, for example, phenylketonuria, a congenital inability to handle a certain dietary constituent, the amino acid phenylalanine, a constituent of many proteins; galactosemia, another inborn biochemical deficiency, the victims of which cannot cope metabolically with galactose, an immediate derivative of milk sugar; and, more common than either, fibrocystic disease of the pancreas, believed to be the symptom of a generalized disorder of mucus-secreting cells. All three are caused by particular genetic defects; but their secondary consequences are manifold and deep-seated. The phenylketonuric baby is on the way to becoming an imbecile. The victim of galactosemia may become blind through cataract and be mentally retarded.

Contrary to popular superstition, many congenital ailments can be prevented or, if not prevented, cured. But in this context prevention and cure have very special meanings.

The phenylketonuric or galactosemic child may be protected from the consequences of his genetic lesion by keeping him on a diet free from phenylalanine in the one case or lactose in the other. This is a most unnatural proceeding, and much easier said than done, but I take it no one would be prepared to argue that it was an unwarrantable interference with the workings of providence. It is not a cure in the usual medical sense because it neither removes nor repairs the underlying congenital deficiency. What it does is to create around the patient a special little world, a microcosm free from phenylalanine or galactose as the case may be, in which the genetic deficiency cannot express itself outwardly.

Now consider the underlying morality of prevention.

We can prevent phenylketonuria by preventing the genetic conjunction responsible for it in the first instance, i.e. by preventing the coming together of an egg cell and a sperm each carrying that same one harmful recessive gene. All but a very small proportion of overt phenylketonurics are the children of parents who are both carriers – carriers, you remember, being the people who inherited the gene from one only of the two gametes that fused at their conception. Carriers greatly outnumber the overtly afflicted. When two carriers of the same gene marry and bear children, one-quarter of their children (on the average) will be normal, one-quarter will be afflicted, and one-half will be carriers like themselves. We shall accomplish our purpose, therefore, if, having identified the carriers – another thing easier said than done, but it *can* be done, and in an increasing number of recessive disorders – we try to discourage them *from marrying each other* by pointing out the likely consequences if they do so. The arithmetic of this is not very alarming. In a typical recessive disease, about one marriage in every five or ten thousand would be discouraged or warned against, and each disappointed party would have between fifty and a hundred other mates to choose from.

If this policy were to be carried out, the overt incidence of a disease like phenylketonuria, in which carriers can be identified, would fall almost to zero between one generation and the next.

Nevertheless the first reaction to such a proposal may be one of outrage. Here is medical officiousness planning yet another insult to human dignity, yet another deprivation of the rights of man. First it was vaccination and then fluoride; if now people are not to be allowed to marry whom they please, why not make a clean job of it and overthrow the Crown or the United States Constitution?

But reflect for a moment. What is being suggested is that a

G

certain small proportion of marriages should be discouraged for genetic reasons, to do our best to avoid bringing into the world children who are biochemically crippled. In all cultures marriages are already prohibited for genetic reasons – the prohibition, for example, of certain degrees of inbreeding (the exact degree varies from one culture or religion to another). It is difficult to see why the prohibition should have arisen to some extent independently in different cultures unless it grew out of the common observation that abnormalities are more common in the children of marriages between close relatives than in children generally. Thus the prohibition of marriage for genetic reasons has an immemorial authority behind it. As to the violation of human dignity entailed by performing tests on engaged couples that are no more complex or offensive than blood tests, let me say only this: if anyone thinks or has ever thought that religion, wealth, or colour are matters that may properly be taken into account when deciding whether or not a certain marriage is a suitable one, then let him not dare to suggest that the genetic welfare of human beings should not be given equal weight.

I think that engaged couples should themselves decide, and I am pretty certain they would be guided by the thought of the welfare of their future children. When it came to be learned, about twenty years ago, that marriages between Rhesus-positive men and Rhesus-negative women might lead to the birth of children afflicted by haemolytic disease, a number of young couples are said to have ended their engagements – needlessly, in most cases, because the dangers were overestimated through not being understood. But that is evidence enough that young people marrying today are not likely to take stand upon some hypothetical right to give birth to defective children, if, by taking thought, they can do otherwise.

The problems I have been discussing illustrate very clearly the way in which scientific evidence bears upon decisions that are not, of course, in themselves scientific. If the termination of a pregnancy is now in question, scientific evidence may tell us that the chances of a defective birth are 100 per cent, 50 per cent, 25 per cent, or perhaps unascertainable. The evidence is highly relevant to the decision, but the decision itself is not a scientific one, and I see no reason why scientists as such should be specially well qualified to make it. The contribution of science is to have enlarged beyond all former bounds the evidence we must take account of before forming our opinions. Today's opinions may not be the same as yesterday's, because they are based on fuller or better evidence. We should quite often have occasion to say 'I used to think that once, but now I have come to hold a rather different opinion.' People who never say as much are either ineffectual or dangerous.

We all nowadays give too much thought to the material blessings or evils that science has brought with it, and too little to its power to liberate us from the confinements of ignorance and superstition.

It may be that the greatest liberation of thought ever achieved by the scientific revolution was to have given mankind the expectation of a future in this world. The idea that the world has at virtually indeterminate future is comparatively new. Much of the philosophic speculation of three hundred years ago was oppressed by the thought that the world had run its course and was coming shortly to an end.[1] 'I was borne in the last Age of the World,' said John Donne, giving it as the 'ordinarily received' opinion that the world had thrice two thousand years to run between its creation and the Second Coming. According to Archbishop Ussher's chronology more than five and a half of those six thousand years had gone by already.[2]

No empirical evidence challenged this dark opinion. There were no new worlds to conquer, for the world was known to be spherical and therefore finite; certainly it was not all known, but the full extent of what was *not* known was known. Outer space did not put into people's minds then, as it does into ours now, the idea of a tremendous endeavour just beginning.

[1] See *The Discovery of Time*, by Stephen Toulmin and June Goodfield (New York, Macmillan, 1965).

[2] 'We are almost the last progeny of the First Men,' said Thomas Burnet.

Moreover, life itself seemed changeless. The world a man saw about him in adult life was much the same as it had been in his own childhood, and he had no reason to think it would change in his own or his children's lifetime. We need not wonder that the promise of the next world was held out to believers as an inducement to put up with the incompleteness and inner pointlessness of this one: the present world was only a staging post on the way to better things. There was a certain awful topicality about Thomas Burnet's description of the world in flames at the end of its long journey from 'a dark chaos to a bright star', for the end of the world might indeed come at any time. And Thomas Browne warned us against the folly and extravagance of raising monuments and tombs intended to last for many centuries. We are living in The Setting Part of Time, he told us: *the Great Mutations of the World are acted: it is too late to be ambitious.*

Science has now made it the ordinarily received opinion that the world has a future reaching beyond the most distant frontiers of the imagination – and that is perhaps why, in spite of all his faults, so many scientists still count Francis Bacon their first and greatest spokesman: we may yet build a New Atlantis. The point is that when Thomas Burnet exhorted us to become 'Adventurers for Another World' *he* meant the next world – but we mean this one.

Lucky Jim[1]

On 30 May 1953 James Watson and Francis Crick published in *Nature* a correct interpretation of the crystalline structure of deoxyribonucleic acid, DNA. It was a great discovery, one which went far beyond merely spelling out the spatial design of a large, complicated, and important molecule. It explained how that molecule could serve genetic purposes – that is to say, how DNA, within the framework of a single common structure, could exist in forms various enough to encode the messages of heredity. It explained how DNA could be stable in a crystalline sense and yet allow for mutability. Above all it explained in principle, at a molecular level, how DNA undergoes its primordial act of reproduction, the making of more DNA exactly like itself. The great thing about their discovery was its completeness, its air of finality. If Watson and Crick had been seen groping towards an answer; if they had published a partly right solution and had been obliged to follow it up with corrections and glosses; if the solution had come out piecemeal instead of in a blaze of understanding: then it would still have been a great episode in biological history, but something more in the common run of things; something splendidly well done, but not done in the grand romantic manner.

The work that ended by making biological sense of the nucleic acids began forty years ago in the shabby laboratories of the Ministry of Health in London. In 1928 Dr Fred Griffith, one of the Ministry's Medical Officers, published in the *Journal of Hygiene* a paper describing strange observations on the behaviour of pneumococci – behaviour which suggested that they could undergo something akin to a transmutation of bacterial species. The pneumococci exist in a variety of genetically different 'types', distinguished one from another by the chemical make-up of their

[1] This article was reprinted with permission from *The New York Review of Books*. Copyright © N.Y. Rev. Inc.

outer sheaths. Griffith injected into mice a mixture of dead pneumococcal cells of one type and living cells of another type, and in
due course he recovered living cells of the type that distinguished
the dead cells in the original mixture. On the face of it, he had
observed a genetic transformation. There was no good reason to
question the results of the experiment. Griffith was a well known
and highly expert bacteriologist whose whole professional life had
been devoted to describing and defining the variant forms of
bacteria, and his experiments (which forestalled the more obvious
objections to the meaning he read into them) were straightforward
and convincing. Griffith, above all an epidemiologist, did not
follow up his work on pneumococcal transformation; nor did he
witness its apotheosis, for in 1941 a bomb fell in Enders Street
which blew up the Ministry's laboratory while he and his close
colleague William Scott were working in it.

The analysis of pneumococcal transformations was carried forward by Martin Dawson and Richard Sia in Columbia University
and by Lionel Alloway at the Rockefeller Institute. Between them
they showed that the transformation could occur during cultivation outside the body, and that the agent responsible for the transformation could pass through a filter fine enough to hold back the
bacteria themselves. These experiments were of great interest to
bacteriologists because they gave a new insight into matters having
to do with the ups and downs of virulence; but most biologists and
geneticists were completely unaware that they were in progress.
The dark ages of DNA came to an end in 1944 with the publication
from the Rockefeller Institute of a paper by Oswald Avery and
his young colleagues, Colin MacLeod and Maclyn McCarty,
which gave very good reasons for supposing that the transforming
agent was 'a highly polymerized and viscous form of sodium
desoxyribonucleate'. This interpretation aroused much resentment,
for many scientists unconsciously deplore the resolution of mys-
teries they have grown up with and have therefore come to love.
It nevertheless withstood all efforts to unseat it. Geneticists mar-
velled at its significance, for the agent that brought about the transformation could be thought of as a naked gene. So very probably

the genes were not proteins after all, and the nucleic acids themselves could no longer be thought of as a sort of skeletal material for the chromosomes.

The new conception was full of difficulties, the most serious being that (compared with the baroque profusion of different kinds of proteins) the nucleic acids seemed too simple in make-up and too little variegated to fulfil a genetic function. These doubts were set at rest by Crick and Watson: the combinatorial variety of the four different bases that enter into the make-up of DNA is more than enough to specify or code for the twenty different kinds of amino acids of which proteins are compounded; more than enough, indeed, to convey the detailed genetic message by which one generation of organisms specifies the inborn constitution of the next. Thanks to the work of Crick and half a dozen others, the form of the genetic code, the scheme of signalling, has now been clarified, and thanks to work to which Watson has made important contributions, the mechanism by which the genetic message is mapped into the structure of a protein is now in outline understood.

It is simply not worth arguing with anyone so obtuse as not to realize that this complex of discoveries is the greatest achievement of science in the twentieth century. I say 'complex of discoveries' because discoveries are not a single species of intellection; they are of many different kinds, and Griffith's and Crick-and-Watson's were as different as they could be. Griffith's was a synthetic discovery, in the philosophic sense of that word. It did not close up a visible gap in natural knowledge, but entered upon territory not until then known to exist. If scientific research had stopped by magic in, say, 1920, our picture of the world would not be *known* to be incomplete for want of it. The elucidation of the structure of DNA was analytical in character. Ever since W. T. Astbury published his first X-ray diffraction photographs we all knew that DNA had a crystalline structure, but until the days of Crick and Watson no one knew what it was. The gap was visible then, and if research had stopped in 1950 it would be visible still; our picture of the world would be known to be imperfect. The importance of

Griffith's discovery was historical only (I do not mean this in a depreciatory sense). He might not have made it; it might not have been made to this very day; but if he had not, then some other, different discovery would have served an equivalent purpose, that is, would in due course have given away the genetic function of DNA. The discovery of the structure of DNA was logically necessary for the further advance of molecular genetics. If Watson and Crick had not made it, someone else would certainly have done so – almost certainly Linus Pauling, and almost certainly very soon. It would have been that same discovery, too; nothing else could take its place.

Watson and Crick (so Watson tells us) were extremely anxious that Pauling should *not* be the first to get there. In one uneasy hour they feared he had done so, but to their very great relief his solution was erroneous, and they celebrated his failure with a toast. Such an admission will shock most laymen: so much, they will feel, for the 'objectivity' of science; so much for all that fine talk about the disinterested search for truth. In my opinion the idea that scientists ought to be indifferent to matters of priority is simply humbug. Scientists are entitled to be proud of their accomplishments, and what accomplishments can they call 'theirs' except the things they have done or thought of first? People who criticize scientists for wanting to enjoy the satisfaction of intellectual ownership are confusing possessiveness with pride of possession. Meanness, secretiveness, and sharp practice are as much despised by scientists as by other decent people in the world of ordinary everyday affairs; nor, in my experience, is generosity less common among them, or less highly esteemed.

It could be said of Watson that, for a man so cheerfully conscious of matters of priority, he is not very generous to his predecessors. The mention of Astbury is perfunctory and of Avery a little condescending. Fred Griffith is not mentioned at all. Yet a paragraph or two would have done it without derogating at all from the splendour of his own achievement. Why did he not make the effort?

It was not lack of generosity, I suggest, but stark insensibility.

These matters belong to scientific history, and the history of science bores most scientists stiff. A great many highly creative scientists (I classify Jim Watson among them) take it quite for granted, though they are usually too polite or too ashamed to say so, that an interest in the history of science is a sign of failing or unawakened powers. It is not good enough to dismiss this as cultural barbarism, a coarse renunciation of one of the glories of humane learning. It points towards something distinctive about scientific learning, and instead of making faces about it we should try to find out why such an attitude is natural and understandable. A scientist's present thoughts and actions are of necessity shaped by what others have done and thought before him; they are the wavefront of a continuous secular process in which The Past does not have a dignified independent existence on its own. Scientific understanding is the integral of a curve of learning; science therefore in some sense comprehends its history within itself. No Fred, no Jim: that is obvious, at least to scientists; and being obvious it is understandable that it should be left unsaid. (I am speaking, of course, about the history of scientific endeavours and accomplishments, not about the history of scientific ideas. Nor do I suggest that the history of science may not be profoundly interesting as history. What I am saying is that it does not often interest the scientist as science.)

Jim Watson ('James' doesn't suit him) majored in Zoology in Chicago and took his PhD in Indiana, aged twenty-two. When he arrived in Cambridge in 1951 there could have been nothing much to distinguish him from any other American 'postdoctoral' in search of experience abroad. By 1953 he was world famous. How much did he owe to luck?

The part played by luck in scientific discovery is greatly over-rated. *Ces hasards ne sont que pour ceux qui jouent bien*, as the saying goes. The paradigm of all lucky accidents in science is the discovery of penicillin – the spore floating in through the window, the exposed culture plate, the halo of bacterial inhibition around the spot on which it fell. What people forget is that Fleming had been *looking* for penicillin, or something like it, since the middle of the First World War. Phenomena such as these will not be appreciated, may not be knowingly observed, except against a background of prior expectations. A good scientist is discovery-prone. (As it happens there *was* an element of blind luck in the discovery of penicillin, though it was unknown to Fleming. Most antibiotics – hundreds are now known – are murderously toxic, because they arrest the growth of bacteria by interfering with metabolic processes of a kind that bacteria have in common with higher organisms. Penicillin is comparatively innocuous because it happens to interfere with a synthetic process peculiar to bacteria, namely the synthesis of a distinctive structural element of the bacterial cell wall.)

I do not think Watson was lucky except in the trite sense in which we are all lucky or unlucky – that there were several branching points in his career at which he might easily have gone off in a direction other than the one he took. At such moments the reasons that steer us one way or another are often trivial or ill thought-out. In England a schoolboy of Watson's precocity and style of genius would probably have been steered towards literary studies. It just so happens that during the 1950s, the first great age of molecular biology, the English Schools of Oxford and particularly of Cambridge produced more than a score of graduates of quite outstanding ability – much more brilliant, inventive, articulate, and dialectically skilful than most young scientists; right up in the Watson class. But Watson had one towering advantage over all of them: in addition to being extremely clever he had something important to be clever *about*. This is an advantage which scientists enjoy over most other people engaged in intellectual pursuits, and they enjoy it at all levels of capability. To be a first-rate scientist it is not necessary (and certainly not sufficient) to be extremely clever, anyhow in a pyrotechnic sense. One of the great social revolutions brought about by scientific research has been the democratization of learning. Anyone who combines strong common sense with an ordinary degree of imaginativeness can become a creative scientist, and a happy one besides, in so far as happiness depends upon being able to develop to the limit of one's abilities.

Lucky or not, Watson was a highly privileged young man. Throughout his formative years he worked first under and then with scientists of great distinction; there were no dark unfathomed laboratories in his career. Almost at once (and before he had done anything to deserve it) he entered the privileged inner circle of scientists among whom information is passed by a sort of beating of tom-toms, while others await the publication of a formal paper in a learned journal. But because it was unpremeditated we can count it to luck that Watson fell in with Francis Crick, who (whatever Watson may have intended) comes out in this book as the dominant figure, a man of very great intellectual powers. By all accounts, including Watson's, each provided the right kind of intellectual environment for the other. In no other form of serious creative activity is there anything equivalent to a collaboration between scientists, which is a subtle and complex business, and a triumph when it comes off, because the skill and performance of a team of equals can be more than the sum of individual capabilities. It was a relationship that did work, and in doing so brought them the utmost credit.

Considered as literature, *The Double Helix* will be classified under Memoirs, Scientific. No other book known to me can be so described. It will be an enormous success, and deserves to be so – a classic in the sense that it will go on being read. As with all good memoirs, a fair amount of it consists of trivialities and idle chatter. Like all good memoirs it has not been emasculated by considerations of good taste. Many of the things Watson says about the people in his story will offend them, but his own artless candour excuses him, for he betrays in himself faults graver than those he professes to discern in others. *The Double Helix* is consistent in literary structure. Watson's gaze is always directed outward. There is no philosophizing or psychologizing to obscure our understanding; Watson displays but does not observe himself. Autobiographies, unlike all other works of literature, are part of their own subject matter. Their lies, if any, are lies *of* their authors but not *about* their authors, who (when discovered in falsehood) merely reveal a truth about themselves, namely that they are liars. Although it sounds a bit too well remembered, Watson's scientific narrative strikes me as perfectly convincing. This is not to say that the apportionments of credits or demerits are necessarily accurate: that is something which cannot be decided in abstraction, but only after the people mentioned in the book have had their say, if they choose to have it. Nor will an intelligent reader suppose that Watson's judgements upon the character, motives, and probity of other people (sometimes apparently shrewd, sometimes obviously petty) are 'true' simply because he himself believes them to be so.

A good many people will read *The Double Helix* for the insight they hope it will bring them into the nature of the creative process in science. It may indeed become a standard case history of the so-called 'hypothetico-deductive' method at work. Hypothesis and inference, feedback and modified hypothesis, the rapid alternation of imaginative and critical episodes of thought – here it can all be seen in motion, and every scientist will recognize the same intellectual structure in the research he does himself. It is characteristic of science at every level, and indeed of most explora-tory or investigative processes in everyday life. No layman who reads this book with any kind of understanding will ever again think of the scientist as a man who cranks a machine of discovery. No beginner in science will henceforward believe that discovery is bound to come his way if only he practises a certain Method, goes through a certain well-defined performance of hand and mind.

Nor, I hope, will anyone go on believing that The Scientist is some definite kind of person. Given the context, one could not plausibly imagine a collection of people more different in origin and education, in manner, manners, appearance, style, and worldly purposes than the men and women who are the characters in this book. Watson himself and Crick and Wilkins, the central figures; Dorothy Crowfoot and poor Rosalind Franklin, the only one of them not then living; Perutz, Kendrew, and Huxley; Todd and Bragg, at that time holder of 'the most prestigious chair in science'; Pauling *père et fils*; Bawden and Pirie, in a momentary appearance; Chargaff; Luria; Mitchison and Griffith (John, not Fred) – they come out larger than life, perhaps, and as different one from another as Caterpillar and Mad Hatter. Watson's child-like vision makes them seem like the creatures of a Wonderland, all at a strange contentious noisy tea party which made room for him because for people like him, at this particular kind of party, there is always room.

On 'The Effecting of All Things Possible'[1]

I

The title of my Address, or if you like its motto, comes from Francis Bacon's *New Atlantis*, published in 1627. The *New Atlantis* was Bacon's dream of what the world might have been, and might still become, if human knowledge were directed towards improving the worldly condition of man. It makes a rather strange impression nowadays, and very few people bother with it who are not interested either in Bacon himself, or in the flux of seventeenth-century opinion or the ideology of Utopias. We shall not read it for its sociological insights, which are non-existent, nor as science fiction, because it has a general air of implausibility; but there is one high poetic fancy in the *New Atlantis* that stays in the mind after all its fancies and inventions have been forgotten. In the New Atlantis, an island kingdom lying in very distant seas, the only commodity of external trade is – *light*: Bacon's own special light, the light of understanding. The Merchants of Light who carry out its business are members of a society or order of philosophers who between them make up (so their spokesman declared) 'the noblest foundation that ever was upon the earth'. 'The end of our foundation', the spokesman went on to say, 'is the knowledge of causes and the secret motions of things; and the enlarging of the bounds of human empire, to the effecting of all things possible.' You will see later on why I chose this motto.

[1] Presidential Address delivered on 3 September 1969, at the Exeter Meeting of the British Association.

II

The purpose of my Address is to draw certain parallels between the spiritual or philosophic condition of thoughtful people in the seventeenth century and in the contemporary world, and to ask why the great philosophic revival that brought comfort and a new kind of understanding to our predecessors has now apparently lost its power to reassure us and cheer us up.

The period of English history that lies roughly between the accession of James I in 1603 and the English Civil War has much in common with the present day.[1] For the historian of ideas, it is a period of questioning and irresolution and despondency; of sermonizing but also of satire; of rival religions competing for allegiance, among them the 'black doctrine of absolute reprobation'; a period during which our human propensity towards hopefulness was clouded over by a sense of inconstancy and decay. Literary historians have spoken of a 'metaphysical shudder',[2] and others of a sense of crisis or of a 'failure of nerve'.[3] Of course, we must not imagine that ordinary people went around with the long sunk-in faces to be expected in the victims of a spiritual deficiency disease. It was philosophic or reflective man who had these misgivings, the man who is all of us some of the time but none of us all of the time, and we may take it that, then as now, the remedy for discomforting thoughts was less often to seek comfort than to abstain from thinking.

Amidst the philosophic gloom of the period I am concerned

[1] See, for example, Herbert Grierson, *Cross Currents in English Literature of the Seventeenth Century* (London, 1929); Basil Willey, *The Seventeenth-Century Background* (London, 1934); G. N. Clark, *The Seventeenth Century*, 2nd ed. (Oxford, 1947); W. Notestein, *The English People on the Eve of Colonization* (New York, 1954); Marjorie Hope Nicolson, *Mountain Gloom and Mountain Glory* (Cornell UP, 1959); Maurice Ashley, *England in the Seventeenth Century*, 3rd ed. (London, 1961); H. R. Trevor-Roper, *Religion, the Reformation and Social Change* (London, 1967).

[2] George Williamson, 'Mutability, decay and seventeenth century melancholy,' *J. Eng. Lit. History*, vol. 2 (1935), pp. 121–50.

[3] Christopher Hill, *Intellectual Origins of the English Revolution* (Oxford, 1965).

with, new voices began to be heard which spoke of hope and of the possibility of a future (a subject I shall refer to later on); which spoke of confidence in human reason, and of what human beings might achieve through an understanding of Nature and a mastery of the physical world. I think there can be no question that, in this country, it was Francis Bacon who started the dawn chorus – the man who first defined the newer purposes of learning and, less successfully, the means by which they might be fulfilled. Human spirits began to rise. To use a good old seventeenth-century metaphor there was a slow change, but ultimately a complete one, in the 'climate of opinion'. It became no longer the thing to mope. In a curious way the Pillars of Hercules – the Fatal Columns guarding the Straits of Gibraltar that make the frontispiece to Bacon's *Great Instauration* – provided the rallying cry of the New Philosophy. Let me quote a great American scholar's, Dr Marjorie Hope Nicolson's,[1] description of how this came about:

> Before Columbus set sail across the Atlantic, the coat of arms of the Royal Family of Spain had been an *impressa*, depicting the Pillars of Hercules, the Straits of Gibraltar, with the motto, *Ne Plus Ultra*. There was 'no more beyond'. It was the glory of Spain that it was the outpost of the world. When Columbus made his discovery, Spanish Royalty thriftily did the only thing necessary: erased the negative, leaving the Pillars of Hercules now bearing the motto, *Plus Ultra*. There was more beyond...

And so *Plus Ultra* became the motto of the New Baconians, and the frontispiece to the *Great Instauration* shows the Pillars of Hercules with ships passing freely to and fro.

One symptom of the new spirit of inquiry was, of course, the foundation of the Royal Society and of sister academies in Italy and France. That story has often been told, and in more than one version, because the parentage of the Royal Society is still in

---

[1] Marjorie Hope Nicolson, 'Two Voices: Science and Literature', *The Rockefeller Review*, vol. no. 3, pp. 1–11, 1963. Reproduced by permission of the Rockefeller UP.

question.[1] We shall be taking altogether too narrow a view of things, however, if we suppose that the great philosophic uncertainties of the seventeenth century were cleared up by the fulfilment of Bacon's ambitions for science. Modern scientific research began earlier than the seventeenth century.[2] The great achievement of the latter half of the seventeenth century was to arrive at a general scheme of belief within which the cultivation of science was seen to be very proper, very useful, and by no means irreligious. This larger conception or purpose, of which science was a principal agency, may be called 'rational humanism' if we are temperamentally in its favour and take our lead from the writings of John Locke, or 'materialistic rationalism' if we are against it and frown disapprovingly over Thomas Hobbes, but neither description is satisfactory, because the new movement had not yet taken on the explicit character of an alternative or even an antidote to religion, which is the sense that 'rational humanism' tends to carry with it today.

However we may describe it, rational humanism became the dominant philosophic influence in human affairs for the next 150 years, and by the end of the eighteenth century the spokesmen of Reason and Enlightenment – men like Adam Ferguson and William Godwin and Condorcet – take completely for granted many of the ideas that had seemed exhilarating and revolutionary in the century before. But over this period an important transformation was taking place. The seventeenth-century doctrine of the *necessity* of reason was slowly giving way to a belief in the *sufficiency* of reason – so illustrating the tendency of many powerful human beliefs to develop into an extreme or radical form before they lose their power to persuade us, and in doing so to create anew many of the evils for which at one time they professed to be the remedy. (It has often been said that rationalism in its more extreme manifestations could only supplant religion by

[1] See, for example, Margery Purver, *The Royal Society: Concept and Creation* (London, 1967) and a number of papers in vol. 23, no. 2 (December 1968) of *Notes and Records of the Royal Society*.

[2] For England in particular, see Christopher Hill, op. cit., note 3 p. 111; F. R. Johnson, *Astronomical Thought in Renaissance Britain* (Baltimore, 1937).

acquiring some of the characteristics of religious belief itself.) Please don't interpret these remarks as any kind of attempt to depreciate the power of reason. I emphasize the distinction between the ideas of the necessity and of the sufficiency of reason as a defence against that mad and self-destructive form of anti-rationalism which seems to declare that because reason is not sufficient, it is not necessary.

Many reflective people nowadays believe that we are back in the kind of intellectual and spiritual turmoil that disturbed the first half of the seventeenth century. Both epochs are marked, not by any characteristic system of beliefs (neither can be called 'The Age of' anything) but by an equally characteristic syndrome of unfixed beliefs; by the emptiness that is left when older doctrines have been found wanting and none has yet been found to take its place. Both epochs have the characteristics of a philosophic interregnum. In the first half of the seventeenth century, the essentially medieval world-picture of Elizabethan England had lost its power to satisfy and bring comfort, just as nowadays the radical materialism traditionally associated with Victorian thinkers seems quite inadequate to remedy our complaints. By a curious inversion of thinking, scholastic reasoning is said to have failed because it discouraged new inquiry, but that was precisely the measure of its success. For that is just what successful, satisfying explanations do: they confer a sense of finality; they remove the incentive to work things out anew. At all events the repudiation of Aristotle and the hegemony of ancient learning, of the scholastic style of reasoning, of the illusion of a Golden Age, is as commonplace in the writings of the seventeenth century as dismissive references to rationalism and materialism in the literature of the past fifty years.

We can draw quite a number of detailed correspondences between the contemporary world and the first forty or fifty years of the seventeenth century, all of them part of a syndrome of dissatisfaction and unbelief; and though we might find reason to cavil at each one of them individually, they add up to an impressive case. Novels and philosophical *belles-lettres* have now an inward-looking character, a deep concern with matters of personal salvation and a struggle to establish the authenticity of personal existence; and we may point to the prevalence of satire and of the Jacobean style of 'realism' on the stage. I shall leave aside the political and economic correspondences between the two epochs,[1] important though they are, and confine myself to analogies that might be described as 'philosophical' in the homely older sense, the sense that has to do with the purpose and conduct of life and with the attempt to answer the simple questions that children ask. Once again we are oppressed by a sense of decay and deterioration, but this time, in part at least, by a fear of the deterioration of the world through technological innovation. Artificial fertilizers and pesticides are undermining our health (we tell ourselves), soil and sea are being poisoned by chemical and radioactive wastes, drugs substitute one kind of disease for another, and modern man is under the influence of stimulants whenever he is not under the influence of sedatives. Once again there is a feeling of despondency and incompleteness, a sense of doubt about the adequacy of man, amounting in all to what a future historian might again describe as a failure of nerve. Intelligent and learned men may again seek comfort in an elevated kind of barminess (but something kind and gentle nevertheless). Mystical syntheses between science and religion, like the Cambridge Neo-Platonism of the mid-seventeenth century, have their counterpart today, perhaps, in the writings and cult of Teilhard de Chardin and in a revival of faith in the Wisdom of the East. Once again there is a rootlessness or ambivalence about philosophical thinking, as if the discovery or rediscovery of the insufficiency of reason had given a paradoxical validity to nonsense, and this gives us a special sympathy for the dilemmas of the seventeenth century. To William Lecky, the

---

[1] England at the time of the Armada was a prosperous country, and it became so again in the reign of Queen Anne; the period I am discussing, however, was marked by a high level of unemployment and a number of major economic slumps, not to mention the English Civil War; moreover the reputation of England abroad sank to a specially low level in the latter part of James Is reign and during the reign of Charles I. This was also the period of the great emigrations to Massachusetts.

great nineteenth-century historian of rationalism, it seemed almost beyond comprehension that witch hunting and witch burning should have persisted far into the seventeenth century, or that Joseph Glanvill should have been equally an advocate of the Royal Society and of belief in witchcraft.[1]

We do not wonder at it now. It no longer seems strange to us that Pascal the geometer who spoke with perfect composure about infinity and the infinitesimal should have been supplanted by Pascal the great cosmophobe who spoke with anguish about the darkness and loneliness of outer space. Discoveries in astronomy and cosmology have always a specially disturbing quality. We remember the dismay of John Donne and Pascal himself and latterly of William Blake. Cosmological discoveries bring with them a feeling of awe but also, for most people, a sense of human diminishment. Our great sidereal adventures today are both elevating and frightening, and may be both at the same time. The launching of a space rocket is (to go back to seventeenth-century language) a tremendous phenomenon. It must have occurred to many who saw pictures of it that the great steel rampart or nave from which the Apollo rockets are launched had the size and shape and grandeur of a cathedral, with Apollo itself in the position of a spire. Like a cathedral it is economically pointless, a shocking waste of public money; but like a cathedral it is also a symbol of aspiration towards higher things.

When we compare the climates of opinion in the seventeenth century and today, we must again remember that cries of despair are not necessarily authentic. There was a strong element of affectation about Jacobean melancholy, and so there is today. Then as now it had tended to become a posture. One of a modern writer's claims to be taken seriously is to castigate complacency and to show up contentment for the shallow and insipid thing that it is assumed to be. But ordinary human beings continue to be vulgarly high spirited. The character we all love best in Boswell is Johnson's old college companion, Mr Oliver Edwards – the man

[1] William Lecky, *The Rise and Influence of Rationalism in Europe* (London, 1865); see especially H. R. Trevor-Roper, op. cit., note 1 p. 111.

who said that he had tried in his time to be a philosopher, but had failed because cheerfulness was always breaking in.

III

I should now like to describe the new style of thinking that led to great revival of spirits in the seventeenth century. It is closely associated with birth of science, of course – of Science with a capital S – and the 'new philosophy' that had been spoken of since the beginning of the century referred to the beginnings of physical science; but (as I said a moment ago) we should be taking too narrow a view of things if we supposed that the establishment or instauration of science made up the whole or even the greater part of it. The new spirit is to be thought of not as scientific, but as something conducive to science; as a movement within which scientific inquiry played a necessary and proper part.

What then were the philosophic elements of the new revival (using 'philosophy' again in its homely sense)?

The seventeenth century was an age of Utopias, though Thomas More's own Utopia was already 100 years old. The Utopias or anti-Utopias we devise today are usually set in the future, partly because the world's surface is either tenanted or known to be empty, partly because we need and assume we have time for the fulfilment of our designs. The old Utopias – Utopia itself, the *New Atlantis*, *Christianopolis*, and the *City of the Sun*[1] – were contemporary societies. Navigators and explorers came upon them accidentally in far-off seas. What is the meaning of the difference? One reason, of course, is that the world then still had room for

[1] J. V. Andreae's *Description of the Republic of Christianopolis* was first published in 1619 (see F. Held, *Christianopolis, an Ideal State of the Seventeenth Century*, Urbana, 1914); Tommaso Campanella published *The City of the Sun* in 1623 (English translation by T. W. Halliday in *Ideal Commonwealths*, London, 1885). There is an extensive literature on Utopian and chiliastic speculation, some of it rather feeble. The following are specially relevant to the idea of progress and of human improvement: J. B. Bury, *The Idea of Progress* (London, 1932); E. L. Tuveson, *Millenium and Utopia* (Berkeley, 1949); Norman Cohn, *The Pursuit of the Millenium* (New York, 1957).

undiscovered principalities, and geographical exploration itself had the symbolic significance we now associate with the great adventures of modern science. Indeed, now that outer space is coming to be our playground, we may again dream of finding ready-made Utopias out there. But this is not the most important reason. The old Utopias were not set in the future because very few people believed that there would *be* a future – an earthly future, I mean; nor was it by any means assumed that the playing-out of earthly time would improve us or increase our capabilities. On the contrary, time was running out, in fulfilment of the great Judaic tradition, and we ourselves were running down.

These thoughts suffuse the philosophic speculation of the seventeenth century until quite near its end. 'I was borne in the last age of the world', said John Donne[1] and Thomas Browne speaks of himself as one whose generation was 'ordained in this setting of time'.[2] The most convincing evidence of the seriousness of this belief is to be found not in familiar literary tags, but in the dull and voluminous writings of those who, like George Hakewill,[3] repudiated the idea of human deterioration and the legend of a golden age, but had no doubt at all about the imminence of the world's end. The apocalyptic forecast was, of course, a source of strength and consolation to those who had no high ambitions for life on earth. The precise form the end of history would take had long been controversial – the New Jerusalem might be founded upon the earth itself or be inaugurated in the souls of men in heaven – but that history would come to an end had hardly been in question. Towards the end of the sixteenth century there had been some uneasy discussion of the idea that the material world might be eternal, but the thought had been a disturbing one, and had been satisfactorily explained away.[4]

[1] In a sermon delivered in Whitehall, 24 February 1625.

[2] In *Hydriotaphia*, his discourse on urn-burial. For a history of the idea of time, see S. Toulmin and J. Goodfield, *The Discovery of Time* (London, 1965).

[3] *An Apologie of the Power and Providence of God* (Oxford, 1627), an answer to Godfrey Goodman's *The Fall of Man* (London, 1616).

[4] See D. C. Allen, 'The degeneration of man and Renaissance pessimism', *Studies in Philology*, vol. 35, pp. 202–27, 1938.

During the seventeenth century this attitude changes. The idea of an end of history is incompatible with a new feeling about the great things human beings might achieve through their own ingenuity and exertions. The idea therefore drops quietly out of the common consciousness. It is not refuted, but merely fades away. It is true that the idea of human deterioration was expressly refuted – in England by George Hakewill but before him by Jean Bodin (by whom Hakewill was greatly influenced) and by Louis de Roy.[1] The refutation of the idea of decay did not carry with it an acceptance of the idea of progress, or anyhow of linear progress: it was a question of recognizing that civilizations or cultures had their ups and downs, and went through a life cycle of degeneration and regeneration – a 'circular kind of progress', Hakewill said.

There were, however, two elements of seventeenth-century thought that imply the idea of progress even if it is not explicitly affirmed. The first was the recognition that the tempo of invention and innovation was speeding up, that the flux of history was becoming denser. In *The City of the Sun* Campanella tells us that 'his age has in it more history within a hundred years than all the world had in four thousand years before it'. He is echoing Peter Ramus:[2] 'We have seen in the space of one age a more plentiful crop of learned men and works than our predecessors saw in the previous fourteen.' By the latter half of the seventeenth century the new concept had sunk in.

The second element in the concept of futurity – in the idea that men might look forward, not only backwards or upwards – is to be found in the breathtaking thought that there was no apparent limit to human inventiveness and ingenuity. It was the notion of a perpetual *Plus Ultra*, that what was already known was only a tiny fraction of what remained to be discovered, so that there would always be more beyond. Bacon published his *Novum*

---

[1] Louis le Roy's remarkable work, addressed to 'all men who thinke that the future belongeth unto them' became known in England through Robert Ashley's translation of 1594 (*Of the Interchangeable Course or Variety of Things*).

[2] Cited by Hakewill, *op cit.* note 3 p. 118.

Organum at the beginning of the remarkable decade between 1620 and 1630, and had singled it out as the greatest obstacle to the growth of understanding, that 'men despair and think things impossible'. 'The human understanding is unquiet', he wrote; 'it cannot stop or rest and still presses onwards, but in vain' – in vain, because our spirits are oppressed by 'the obscurity of nature, the shortness of life, the deceitfulness of the senses, the infirmity of judgement, the difficulty of experiment, and the like'. 'I am now therefore to speak of hope', he goes on to say, in a passage that sounds like the trumpet calls in *Fidelio*. The hope he held out was of a rebirth of learning, and with it the realization that if men would only concentrate and direct their faculties, 'there is no difficulty that might not be overcome'. 'The process of Art is indefinite', wrote Henry Power, 'and who can set a *Plus Ultra* to her endeavours?'.[1] There is a mood of exultation and glory about this new belief in human capability and the future in which it might unfold. With Thomas Hobbes 'glorying' becomes almost a technical term: 'Joy, arising from imagination of a man's own power and ability, is that exultation of mind called glorying', he says, in *Leviathan*, and in another passage he speaks of a 'perseverance of delight in the continual and indefatigable generation of knowledge'.

It does not take a specially refined sensibility to see how exciting and exhilarating these new notions must have been. During the eighteenth century, of course, everybody sobers up. The idea of progress is taken for granted – but in some sense it gets out of hand, for not only will human inventions improve without limit, but so also (it is argued, though not very clearly) will human beings. It is interesting to compare the exhilaration of the seventeenth century with, say, William Godwin's magisterial tone of voice as the eighteenth century draws to an end. 'The extent of our progress in the cultivation of human knowledge is unlimited.

[1] *Experimental Philosophy*, Preface (London, 1664); cf. John Ray, *The Wisdom of God Manifested in the Works of Creation*, (2nd ed. pp. 164–5; 1st ed., London, 1691).

Hence it follows ... that human inventions are susceptible of perpetual improvement.'[1]

> Can we arrest the progress of the inquiring mind? If we can, it must be by the most unmitigated despotism. Intellect has a perpetual tendency to proceed. It cannot be held back but by a power that counteracts its genuine tendency through every moment of its existence. Tyrannical and sanguinary must be the measures employed for this purpose. Miserable and disgustful must be the scene they produce.

The seventeenth century had begun with the assumption that a powerful force would be needed to put the inventive faculty into motion; by the end of the eighteenth century it is assumed that only the application of an equally powerful force could possibly slow it down.

Before going on, it is worth asking if this conception is still acceptable – that the growth of knowledge and know-how has no intrinsic limit. We have now grown used to the idea that most ordinary or natural growth processes (the growth of organisms or populations of organisms or, for example, of cities) is not merely limited, but self-limited, i.e. is slowed down and eventually brought to a standstill as a *consequence of the act of growth itself.* For one reason or another, but always for some reason, organisms cannot grow indefinitely, just as beyond a certain level of size or density a population defeats its own capacity for further growth. May not the body of knowledge also become unmanageably large, or reach such a degree of complexity that it is beyond the comprehension of the human brain? To both these questions I think the answer is 'No'. The proliferation of recorded knowledge and the seizing-up of communications pose technological problems for which technical solutions can and are being found. As to the idea that knowledge may transcend the power of the human brain: in a sense it has long done so. No one can 'understand' a radio-set or

---

[1] William Godwin, *Enquiry Concerning Political Justice* (3rd ed., London, 1797; 1st ed., 1793). Cf. Adam Ferguson, *An Essay on the History of Civil Society* (Edinburgh, 1767).

automobile in the sense of having an effective grasp of more than a fraction of the hundred technologies that enter into their manufacture. But we must not forget the additiveness of human capabilities. We work through consortia of intelligences, past as well as present. We might, of course, blow ourselves up or devise an unconditionally lethal virus, but we don't *have* to. Nothing of the kind is necessarily entailed by the growth of knowledge and understanding. I do not believe that there is any intrinsic limitation upon our ability to answer the questions that belong to the domain of natural knowledge and fall therefore within the agenda of scientific inquiry.

IV

The repudiation of the concept of decay, the beginnings of a sense of the future, an affirmation of the dignity and worthiness of secular learning, the idea that human capabilities might have no upper limit, an exultant recognition of the capabilities of man – these were the seventeenth century's antidote to despondency. You may wonder why I have said nothing about the promulgation of the experimental method in science as one of the decisive intellectual movements of the day. My defence is that the origin of the experimental method has been the subject of a traditional misunderstanding, the effect of reading into the older usages of 'experiment' the very professional meaning we attach to that word today. Bacon is best described as an advocate of the *experiential* method in science – of the belief that natural knowledge was to be acquired not from authority, however venerable, nor by syllogistic exercises, however subtle, but by paying attention to the evidence of the senses, evidence from which (he believed) all deception and illusion could be stripped away. Bacon's writings form one of the roots of the English tradition of philosophic empiricism, of which the greatest spokesman was John Locke. The unique contribution of science to empirical thought lay in the idea that experience could be *stretched* in such a way as to make nature yield up information which we should otherwise have been unaware of. The word 'experiment' as it was used until the nineteenth century stood for the concept of stretched or deliberately contrived experience; for the belief that we might make nature perform according to a scenario of our own choosing instead of merely watching her own artless improvisations. An 'experiment' today is not something that merely enlarges our sensory experience. It is a critical operation of some kind which discriminates between hypotheses and therefore gives a specific direction to the flow of thought. Bacon's championship of the idea of experimentation was part of a greater intellectual movement which had a special manifestation in science without being distinctively scientific. His reputation should not, and fortunately need not rest on his being the founder of the 'experimental method' in the modern sense.[1]

Let us return to the contemporary world and discuss our misgivings about the way things are going now. No one need suppose that our present philosophic situation is unique in its character and gravity. It was partly to dispel such an illusion that I have been moving back and forth between the seventeenth century and the present day. Moods of complacency and discontent have succeeded each other during the past 400 or 500 years of European history, and our present mood of self-questioning does not represent a new and startled awareness that civilization is coming to an end. On the contrary, the existence of these doubts is probably our best assurance that civilization will continue.

Many of the ingredients of the seventeenth-century antidote to melancholy have lost their power to bring peace of mind today, and have become a source of anxiety in themselves. Consider the tempo of innovation. In the post-Renaissance world the feeling that inventiveness was increasing and that the whole world was on the move did much to dispel the myth of deterioration and give people confidence in human capability. Nevertheless the tempo

[1] See my Jayne Lectures, *Induction and Intuition in Scientific Thought* (Philadelphia and London, 1969). The idea of 'stretched experience' and of the experimenter as the 'archmaster' who 'completes experience', comes from John Dee's *Mathematical Preface* to Henry Billingsley's English translation of Euclid (London, 1570).

was a pretty slow one, and technical innovation had little influence on the character of common life. A man grew up and grew old in what was still essentially the world of his childhood; it had been his father's world and it would be his children's too. Today the world changes so quickly that in growing up we take leave not just of youth but of the world we were young in. I suppose we all realize the degree to which fear and resentment of what is new is really a lament for the memories of our childhood. Dear old steam trains, we say to ourselves, but nasty diesel engines; trusty old telegraph poles but horrid pylons. Telegraph poles, as the Poet Laureate told us a good many years ago,[1] are something of a test case. Anyone who has spent part of his childhood in the country-side can remember looking up through the telegraph wires at a clouded sky and discerning the revolution of the world, or will have listened, ear to post, to the murmur of interminable con-versations. For some people even the smell of telegraph poles is nostalgic, though creosote has a pretty technological smell. Tele-graph poles have been assimilated into the common consciousness, and one day pylons will be, too. When the pylons are dismantled and the cables finally go underground, people will think again of those majestic catenary curves, and remind each other of how steel giants once marched across the countryside in dead silence and in single file. (What is wrong with pylons is that most of them are ugly. If only the energy spent in denouncing them had been directed towards improving their appearance, they could have been made as beautiful, even as majestic, as towers or bridges are allowed to be, and need not have looked incongruous in the countryside.)

When Bacon described himself as a trumpeter of the new philosophy, the message he proclaimed was of the virtue and dignity of scientific learning and of its power to make the world a better place to live in. I am continually surprised by the super-ficiality of the reasons which have led people to question those beliefs today. Many different elements enter into the movement to depreciate the services to mankind of science and technology. I

[1] C. Day Lewis, *A Hope for Poetry*, p. 107 (London, 1934).

have just mentioned one of them, the tempo of innovation when measured against the span of life. We wring our hands over the miscarriages of technology and take its benefactions for granted. We are dismayed by air pollution but not proportionately cheered up by, say, the virtual abolition of poliomyelitis. (Nearly 5,000 cases of poliomyelitis were recorded in England and Wales in 1957. In 1967 there were less than 30.) There is a tendency, even a perverse willingness to suppose that the despoliation sometimes produced by technology is an inevitable and irremediable process, a trampling down of Nature by the big machine. Of course it is nothing of the kind. The deterioration of the environment produced by technology is a technological problem for which technology has found, is finding, and will continue to find solutions. There is, of course, a sense in which science and technology can be arraigned for devising new instruments of warfare, but another and more important sense in which it is the height of folly to blame the weapon for the crime. I would rather put it this way: that in the mangement of our affairs we have too often been bad workmen, and like all bad workmen we blame our tools. I am all in favour of a vigorously critical attitude towards technological innovation: we should scrutinize all attempts to improve our condition and make sure that they do not in reality do us harm; but there is all the difference in the world between informed and energetic criticism and a drooping despondency that offers no remedy for the abuses it bewails.

Superimposed on all particular causes of complaint is a more general cause of dissatisfaction. Bacon's belief in the cultivation of science for the 'merit and emolument of life' has always been repugnant to those who have taken it for granted that comfort and prosperity imply spiritual impoverishment. But the real trouble nowadays has very little to do with material prosperity or technology or with our misgivings about the power of research and learning generally to make the world a better place. The real trouble is our acute sense of human failure and mismanagement, a new and specially oppressive sense of the inadequacy of man. So much was hoped of us, particularly in the eighteenth century. We were going to improve, weren't we? – and for some reason which was never made clear to us we were going to grow in moral stature as well as in general capability. Our school reports were going to get better term by term. Unfortunately they haven't done so. Every folly, every enormity that we look back on with repugnance can find its equivalent in contemporary life. Once again our intellectuals have failed us; there is a general air of misanthropy and self-contempt, of protest, but not of affirmation. There is a peculiar selfishness about modern philosophic speculation (using 'philosophy' here again in its homely or domestic sense). The philosophic universe has contracted into a neighbourhood, a suburbia of personal relationships. It is as if the classical formula of self-interest, 'I'm all right, Jack', was seeking a new context in our private, inner, world.

We can obviously do better than this, and there is just one consideration that might help to take the sting out of our self-reproaches. In the melancholy reflections of the post-Renaissance era it was taken for granted that the poor old world was superannuated, that history had all but run its course and was soon coming to an end. The brave spirits who inaugurated the new science dared to believe that it was *not* too late to be ambitious, but now we must try to understand that it is a bit too early to expect our grander ambitions to be fulfilled. Today we are conscious that human history is only just beginning. There has always been room for improvement; now we know that there is time for improvement, too. For all their intelligence and dexterity – qualities we have always attached great importance to – the higher primates (monkeys, apes and men) have not been very successful. Human beings have a history of more than 500,000 years. Only during the past 5,000 years or thereabouts have human beings won a reward for their special capabilities; only during the past 500 years or so have they begun to be, in the biological sense, a success. If we imagine the evolution of living organisms compressed into one year of cosmic time, then the evolution of man has occupied a day. Only during the past 10 to 15 minutes of the human day has our life on earth been anything but precarious.

Until then we might have gone under altogether or, more likely, have survived as a biological curiosity; as a patchwork of local communities only just holding their own in a bewildering and hostile world. Only during this past 15 minutes (for reasons I shall not go into though I think they can be technically explained) has there been progress, though, of course, it doesn't amount to very much. We cannot point to a single definitive solution of any one of the problems that confront us – political, economic, social or moral, i.e. having to do with the conduct of life. We are still beginners, and for that reason may hope to improve. To deride the hope of progress is the ultimate fatuity, the last word in poverty of spirit and meanness of mind. There is no need to be dismayed by the fact that we cannot yet envisage a definitive solution of our problems, a resting-place beyond which we need not try to go. Because he likened life to a race,[1] and defined felicity as the state of mind of those in the front of it, Thomas Hobbes has always been thought of as the arch materialist, the first man to uphold go-getting as a creed. But that is a travesty of Hobbes's opinion. He was a go-getter in a sense, but it was the going, not the getting he extolled. The race had no finishing post as Hobbes conceived it. The great thing about the race was to be in it, to be a contestant in the attempt to make the world a better place, and it was a spiritual death he had in mind when he said that to forsake the course is to die. 'There is no such thing as perpetual tranquillity of mind while we live here', he told us in *Leviathan*, 'because life itself is but motion and can never be without desire, or without fear, no more than without sense'; 'there can be no contentment but in proceeding'. I agree.

[1] This simile occurs more than once in Hobbes; the passage I have in mind is from his *Human Nature* (London, 1650).

Index